# FOUL FACTS

# OUR WORLD

# THE AWFUL TRUTH!

Amber Grayson, Jamie Stokes and friends

Illustrations by

Mike Phillips

*p*

This is a Parragon Book
First published in 2001

Parragon
Queen Street House
4 Queen Street
Bath BA1 1HE, UK

This edition copyright © Parragon 2001

Produced by Magpie Books, an imprint of
Constable & Robinson Ltd, London

ISBN 0-75255-304-6

A copy of the British Library Cataloguing-in-Publication Data
is available from the British Library

Printed and bound in China

# CONTENTS

# INTRODUCTION

Hello, my name is Albert and I am a collector. I don't collect boring stuff like stamps or coins or beetles (unless I'm hungry!), I collect wacky facts and weirdness. In my travels all over the world (and beyond) I have found some of the strangest, most brain-bending stories that have ever been told. Welcome to my museum!

Step inside to see the Wacky Ancient World. Find out what really happened to Atlantis and why it's not a good idea to dig up a cursed mummy!

Move on to the Wacky Natural World where you can marvel at the wave that was almost half-a-mile high and find out why you should pay attention the next time you hear a dog howling.

In Wacky Superstitions you will find some of the craziest, weirdest beliefs ever heard. Did you know it was bad luck to cut your toenails on a Sunday? You have been warned!

Right next door is the Wacky World of Animals where you can meet the cat that hitchhiked across America. Nearby are Wacky Creepy Crawlies and the Wacky World of Fish. You'll never look at cod and chips in the same way again!

If you still want more there's plenty of strangeness in Wacky Plants. Can you imagine a carrot three times as tall as you, and what about the rabbit that's after it?!

In the Wacky World of Humans you can find out just how mad people can be. Would you believe that eating a bible could kill you?

There's plenty more barely believable stuff in Wacky Records. How would you like a car with a swimming pool and bed in it? Meet the man who has one!

If your stomach is strong enough, take a look at Wacky Food. Would you be tempted by a 30-mile-long hotdog, and where would you find enough ketchup to put on it?!

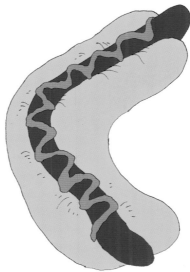

Things start to get very weird and scary if you venture into Wacky Dreams and Wacky Monsters. What kind of creatures lurk in the world's deepest lakes, and is it safe to go swimming in them?!

All aboard for Wacky Ship Stories next. Do you dare venture into the mysteries of the deadly Bermuda Triangle?

For the ultimate in wacky weirdness enter the Wacky Supernatural World. Who are the mysterious men in black, what goes on at Roswell air base, and why do aliens think that Star Wars is so funny?

Still want more? Last up comes Wacky Double Takes and the Final Wacky Assortment. Find out about the world's weirdest twins and why it might be a good idea to eat wood lice!

# WACKY ANCIENT WORLD

### Lost It!
The Lost City of Atlantis was a rich and beautiful island city with fertile soil, inhabited by artistic and educated people. And then it sank. Or did it? Was Atlantis destroyed by volcanic eruptions and a tidal wave or did it exist only in people's imaginations?

## Vision of Atlantis

Earlier this century, Edgar Cayce, an American psychic (a person with strange mind powers), claimed he had visited Atlantis in a dream. According to Cayce, the people of Atlantis were far more advanced than any other civilization of the time, and apparently they even had televisions and X-rays (but did they have a McDonalds?!).

## Foundations Found

Cayce dreamed that Atlantis was built near a place which is now called Bimini, in the Bahamas, and the psychic predicted that in the year 1968 part of the city would be revealed. Oddly enough, scuba divers did find what look like massive foundation stones under the sea near Bimini in the year that Cayce said they would.

Unfortunately, we still don't know if it was the legendary City of Atlantis or just any old city under the sea.

## Tutankhamun's Tomb

Tutankhamun was the boy pharaoh who ruled Egypt over 3,200 years ago. He died when he was 18 years old. His tomb and its treasures remained undisturbed for over 32 centuries until Howard Carter and Lord Caernarvon found it in 1922.

## Find of the Century

It was very unusual to find an untouched tomb – most of the other pharaohs' tombs had been robbed of their treasures over the centuries.

But Tutankhamun's four-roomed tomb contained more than 5,000 beautiful items, including musical instruments, thrones and several statues and gold masks, as well as the mummified body of Tutankhamun. It was hailed as the archaeological find of the century.

## The Mummy's Curse

Not what your mother does when you walk all over her kitchen floor with your muddy shoes, but something a little more mysterious and sinister!

It seems that several people who were connected with the opening of Tutankhamun's tomb lost their lives shortly afterwards. Howard Carter's pet canary, bought just after the discovery of the tomb, was killed by a king cobra, a snake which was often used by the ancient Egyptians to represent the pharaoh. When the canary died, witnesses say that it made an "almost human" cry. This spread rumors of a terrible curse on the tomb and all those who'd been responsible for its discovery.

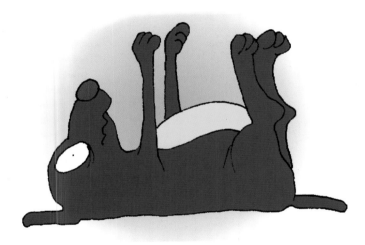

## Germ Warfare

Much later, in 1962, Dr Ezzedin Taha came up with the idea that all the deaths were caused by infection from some ancient lethal bacteria which had been sealed into the tomb to protect it from robbers. But just a few days after explaining his theory, Dr Taha died in a mysterious car accident!

## Off the Hook

Howard Carter seemed to have got away with it, yet he was the man who was responsible for this fantastic archaeological discovery. Carter spent a further ten years working in the tomb and listing its contents, and died of natural causes at the age of 65.

## Bitten to Death

A few weeks after the canary died, Lord Caernarvon became sick from an infected insect bite and, to the shock of all those around him, died from the infection. The weird thing was that at his moment of death all the lights in Cairo went out for no apparent reason! Even more strange, back home in England, Lord Caernarvon's dog dropped down dead at the same time as his master.

Another archaeologist and a friend of Howard Carter were next on the list. George Bendite and Arthur Mace came to visit the tomb and not long after, both men died suddenly.

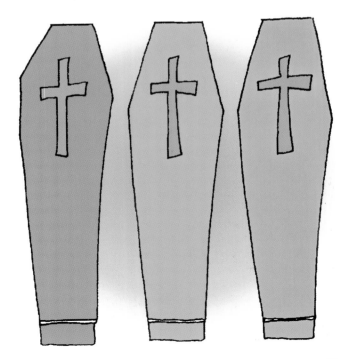

## Great Pyramid

The Great Pyramid at Giza is the oldest of the Seven Wonders of the World and the only one still standing. It's an incredible feat of engineering because it's made from over 2 million blocks of stone fitted together without mortar – in other words, they aren't stuck to each other!

The Great Pyramid was built as a tomb for one of the pharaohs, King Khufu (or "Cheops" as the Greeks called him), and when it was built it was the tallest building in the world. It covers an immense area – bigger than nine football pitches. It's no wonder that experts estimate that it probably took as many as 100,000 men over 20 years to build it.

## Egyptian Sphinx

The Sphinx has the head of a human, the body of a lion and guards the pyramids at Giza, Egypt. It was built about 4,500 years ago (although there are people who think that it might be even older than that). The face of the Sphinx is over 4 meters (14 ft) wide and is said to be modeled on Khafre, one of the pharaohs of Egypt who ruled from 2556–2530 BC.

## Rock Solid

The statue is 20 meters (66 ft) high and 73 meters (241 ft) long and, except for its legs and paws, is carved from one piece of rock. It is said to have secret underground chambers, which have supposedly been discovered quite recently.

## Martian Face

There are many theories about how the Sphinx was built, and there are people who think that because it's such a colossal statue, extraterrestrials might have had something to do with it. This is backed up by the recent discovery of the face of *another* Sphinx on the surface of the planet Mars, which bears a very close resemblance to the Egyptian one.

WHERE ARE THESE NINE FOOTBALL PITCHES THEN ?

## The Sphinx's Riddle

According to legend, there is a riddle which the Sphinx asks anyone who walks by:

*What animal has one voice, but four feet in the morning, two feet at midday and three feet at sunset?*

Oh, by the way, if you're ever asked this riddle, and you get it wrong, you'll suffer instant death (according to Greek mythology), so here's the answer:

*Man.*

Man crawls on all fours as a baby, in the "morning" of his life, walks on two feet as a man, in the "afternoon" of his life, and uses a stick to help him walk in the "sunset" of his life. Phew! That reply should save anyone from a fate worse than death!

# Whodunnit?

In 1996 a large stone pyramid was found under the Pacific Ocean near the island of Yonaguni in Japan. The pyramid does not look as if it was formed naturally, but if it *was* built by humans it had to have been made before sea levels rose in that area, which was at least 10,000 years ago (5,000 years before the Egyptians built theirs). Have those extraterrestrials been at it again, or was there a *really* ancient civilization which we haven't yet discovered?

# Tunnel Vision

In 530 BC, Epalinus of Megara became the first person to build a tunnel through a mountain, starting from both ends at once and meeting in the middle. Epalinus was Greek, so we know he was a clever clogs, but did he have any help, we wonder?

# El Dorado

Not some tacky television soap but a legendary lake full of gold in Central America. It was thought to be Lake Guatavita in Colombia, which sits 2,300 meters (9,000 ft) above sea-level in the crater of an extinct volcano. In the sixteenth century, one of the conquering Spaniards, Perez de Quesada, tried to drain the lake to get at the gold. It was too big a task and he gave up after finding only a few bits of jewelry. In 1912 the lake was drained completely by a wealthy Colombian. Some gold was found but nowhere near enough to cover the expense of draining the lake. So either the native Indians made up the story to fool the Spaniards, or there is a lake full of gold still waiting to be discovered.

## Stone Circles

Avebury is the largest stone circle in Great Britain, but Stonehenge is probably the most famous and the most mysterious.

Some sites like Stonehenge were in use for around 1,000 years so it's incredible that we still don't know exactly *why* these circles were made.

Some of the tools used in the building of the stone circles, such as pieces of flint and picks made from antlers, have been found. But there is no trace of litter on the sites (no crisp packets and coke cans of course, but not even animal bones from food eaten).

So were they built by priests for worship or used as landing sites by extraterrestrials?

IT'S FINISHED, BUT WHAT ON EARTH ARE WE GOING TO DO WITH IT?

## Longest Day

What we do know is that Stonehenge was a sort of giant stone calendar whose huge stones show the movements of the Sun and Moon over hundreds of years. Modern-day Druids perform ceremonies celebrating the summer solstice – the longest day – at the stone circles. The Druids existed in Britain before the Roman Conquest. Julius Caesar described them as men of great learning, given to discussions of stars and their movements, the size of the universe and of the Earth. But some of their activities were less refined. Their rituals included human sacrifices in which living men were set on fire. Although we know all this about what the Druids got up to, there is no hard evidence to link the Druids with stone circles.

# WACKY ANCIENT WORLD QUIZ

Did you enjoy that little collection? Take a minute to look at this little quiz. See how many of the questions you can answer and then try it out on your friends. I bet they won't believe that any of the answers can be right. You and I know different though, don't we!

1. **Edgar Cayce believed that the people of Atlantis had:**

a  Television sets?

b  McDonalds restaurants?

c  Leaky baths?

TOOT AND COME IN

2. **Tutankhamun was:**

a  An archaeologist?

b  A pharaoh of ancient Egypt?

c  The sign on a friendly Egyptian door (toot-and-come-in)!?

**3.** The Great Pyramid at Giza was constructed from:

a  Ten thousand blocks of stone?

b  Two million blocks of stone?

c  Two million blocks of cheese?

**4.** The Sphinx has:

a  The body of a lion and the head of a man?

b  The body of a man and the head of a lion?

c  The body of a lion and the head of Dumbo the Elephant?

**5.** The legendary lake of El Dorado was said to be full of:

a  Water?

b  Gold?

c  Custard?

**6.** The largest stone circle in Great Britain is:

a  Stonehenge?

b  Avebury?

c  A traffic roundabout in Birmingham?

Answers on page 172.

# WACKY NATURAL WORLD

YOU'RE IT!

## Astonishing Stones

There are some weird stones that move by themselves on a dried-up lake on the edge of California's Death Valley. The tracks made by the stones can be seen clearly, and some are more than 80 meters (264 ft) long. Nobody knows what shifts them and no one has seen them move.

If they were pushed along by the wind they would all move in the same direction, but each stone has its own route, and makes individual patterns in the silt of the lake floor. The moving stones were thought to be a hoax, but research has shown that even when there are no people about to move the stones, they still move.

Did you know that the word "hoax" comes from the saying *hocus pocus*, which itself came from the name of a sorcerer (wizard) called Ochus Bochus.

16

# Great Balls of Fire!

Moving balls of fire have been around for a long time. Long ago people thought that they were to do with evil spirits or the devil. Now it has been discovered that these fireballs are caused by weather conditions and are known as *ball lightning*. This orange, glowing ball makes a crackling, whirring sound as it shoots along the ground. It is very hot and can burn anything that gets in its way. Sometimes it disappears in an explosion, but ball lightning has been known to find its way into a bathroom – only to disappear down the plughole. It has even traveled through a plate glass window and left a hole with the edges sealed by its heat.

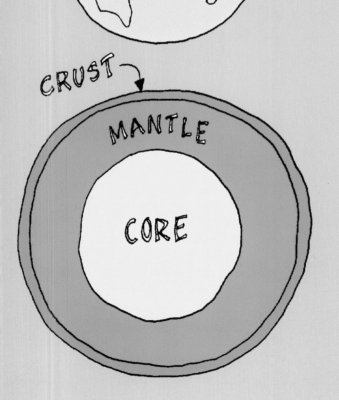

## Earthquakes

*What is an Earthquake?*
We all know, even if we haven't experienced it ourselves, that during an earthquake the ground shakes, lots of buildings fall down and sometimes cracks appear in the Earth. But what makes an earthquake *happen*?

*Crusty Earth*
Planet Earth is covered with a thick crust. The crust is cracked into more than a dozen sections – a bit like the shell of a boiled egg when you tap it with a spoon. These cracked sections are called *crustal plates,* or *plates*.

*Plate Walking*
We walk on top of crustal plates; they make up the mountains and valleys and ocean floors and islands and continents. When we're driving along the road, walking down the sidewalk, playing in the playground, we're doing so on the Earth's plates.

*Floating Plates*
Underneath these plates is another layer of rock, which in some places is so hot that the rock has gone all runny so the plates are floating on top of it. This second layer is called the *mantle* and underneath that, at the centre of the Earth, is the *core*.

17

## Moving Mantle

In the mantle, the liquid or *molten* part is always moving, and this causes the plates to move away from or towards each other. When the plates rub against each other or collide with each other, an earthquake happens.

## No Fault of Their Own

Earthquakes do much more than shake the ground. In some places the ground splits apart, along a line called a *fault*. In other places, where there is loose rock and soil, ground vibration can start landslides.

## Chip Off the Old Block

*Shock waves* from an earthquake caused ice to fall from one of the peaks of Mount Huascaran in Peru. The rocks and debris formed a landslide and a surging river of mud wiped out the town of Yungay in the valley below, killing 18,000 people. The only thing left visible was the roof of a monastery with a statue of Jesus poking out of the top.

## Water Wall

Another terrifying by-product of an earthquake (as if they need one!) is a *tsunami*. This is a huge wave, again caused by an earthquake, but this time under the sea bed. In Alaska, in 1958, shock waves created a massive wall of water 524 meters (1,720 ft) high, which moved at 160 km (100 miles) per hour, as fast as a speeding car. In fact, tsunamis actually slow down when they reach the coastline but these giant sea waves can hurtle across deep oceans at up to 600 km (400 miles) an hour! Surf 'em if you dare!

## In For a Shock

San Francisco is a rich, thriving city on the west coast of North America. People live there and are still building houses, shops and office blocks, but San Francisco sits on one the greatest earthquake zones (faults) in the world.

## Shake Up

There have been quite a few earthquakes there – in 1906 a 'quake destroyed an area 50 km (30 miles) around its center. Rumor has it that in the not too distant future there is going to be a massive earth disturbance in California. Children have earthquake drills at school, but, unsurprisingly, people are hoping that it will never happen.

## Sounding the Alarm

In the recent earthquake in Turkey in 1999, dogs howled like wolves for two days before the earthquake struck. No one took any notice and many of the buildings, which were badly constructed, collapsed like stacks of cards, killing over 17,000 people.

## Massive Destruction

The highest recorded death toll caused by an earthquake was in China in 1556 when at least 830,000 people died. China also holds the record for the greatest number of deaths from an earthquake in recent times. In 1976, a huge 'quake caused the deaths of 250,000 people.

## Deep Space

The Rift Valley in Africa was created by repeated earthquakes over thousands of years, which opened up the ground to form an enormous trench. The trench is so deep and wide that it can be seen from the moon!

## Volcanoes

Volcanoes sit on the site of a weak point in the Earth's crust. When the pressure builds up underneath them, they erupt. Some volcanoes have *chimneys* – not the sort you expect Father Christmas to come down unless he's wearing his extra-thick asbestos suit, but a narrow funnel in the peak of the volcano. Boiling-hot sticky stuff called *magma* spews out of this hole in the middle or out of *fissures* which run down the sides of the volcano. Fissures aren't people who supply your local fish shop with cod. They are cracks in the Earth's crust. The magma oozes out of the fissures or erupts from the chimney and pours down the mountain, creating a river of lava or a *lava flow*. The longest lava flow was 70 km (43.5 miles) long and came from an eruption of a volcano called Laki (un-Laki for some!) in Iceland.

When a volcano is *extinct* it means that no one expects it to erupt again, but the word "extinct" isn't often used about volcanoes – you just never know when they might pop off! Most inactive volcanoes are called *dormant*, which means that they aren't doing anything at the moment but they might do!

There are around 1,300 active volcanoes in the world and about 500 of these are active at any one time. The most active volcano is Killauea in Hawaii and the highest active volcano is Ojos de Salado which is 6,885 meters (22,720 ft) high and is in the South American Andes.

## Island Magic

Many active volcanoes are under the sea and a lot goes on that we don't know about. Sometimes a *submarine eruption*, as it is known, causes a new island to be formed. Surtsey Island, off the coast of Iceland, was formed between 1963 and 1966 but the newest island is near Tonga. It is 5 hectares (12 acres) wide and was formed in June 1995. You wouldn't want to go and live there just yet because it might grow bigger under your feet!

## Short-changed

As well as making new islands, volcanic eruptions can cause existing islands to be reduced in size. An explosive eruption of the Tambora volcano on Sumbawa in Indonesia caused the height of the island to drop by 1,250 meters (4,100 ft).

## Preserved State

Pompeii was a Roman town which sat under the volcano Vesuvius. When Vesuvius erupted in AD 79, 20,000 people were trapped and killed by the hot ash as they went about their daily business. The whole town was preserved by the ash and the places where the bodies of people and animals have decayed over the years have made hollow shapes or molds and can be seen today.

### Mountain Out of a Molehill

In 1943 a small mound appeared in a field in Mexico. In a month it was 150 meters (495 ft) high. Now Paricutin, as it has been named, is a 2,700-meter-high (8,910 ft) volcano, and needless to say, the farmer's field is no longer.

### Great Crater

A *caldera* is a huge volcanic crater. Sometimes these become filled with rainwater which forms a lake inside the volcano. The biggest crater in the world is Toba in Sumatra, Indonesia, and it covers an area of 1,775 sq km (685 sq miles).

### Recent Explosion

In May 1980, in Washington State, USA, Mount St Helens popped off spectacularly. Eruptions went on for four days, blowing a horseshoe-shaped crater and showering the surrounding area with suffocating ash and gases which burst out sideways from the mountain. The glacier at the top of the mountain melted, causing flooding in the valleys below, but because the volcano had been closely monitored before the eruption only 85 people were killed. Many of these were sightseers who didn't expect the volcano to go off with such force and were suffocated by the gas and smoke as they sat in their cars.

## Big Bang

The loudest explosion was probably New Zealand's Mount Taupu. It went off less than a couple of thousand years ago and threw out 30 billion tonnes of rock and ash which were hurled into the air at around 700 km (400 miles) per hour.

## Old as the Hills

Who would have thought it? The tors on Dartmoor in the UK are really the remains of ancient volcanoes that exploded and eroded over millions of years, until they became relatively insignificant bumps in the ground.

## Young Rock

The highest mountain range is the Himalayas on the borders of India and China. They've only been around for 65 million years or so, and haven't had time to crumble. In fact, if you see a big mountain it's got to be a young 'un. The very, very old ones are much smaller.

## Exploding Castle

Edinburgh Castle in Scotland is built on the site of an old volcano. No need to worry though, it last popped off one thousand million years ago, and now you'd probably get more explosions out of a haggis than you would out of that volcano.

## Good Crack

In 1883, the biggest volcanic eruption in recorded history occurred when the island of Krakatoa exploded with about 26 times the strength of the biggest nuclear bomb test. It caused a tidal wave (also called a "tsunami") which killed more than 36,000 people and dust from the explosion fell over 5,000 km (3,125 miles) away.

## Gentle Giant

Aconcagua in Chile is the highest volcano on this planet. It is a massive 6,690 meters (22,834 ft) but luckily for the people who live beneath it, it has been dormant for years and shows no signs of doing anything in the near future.

# Geysers and Hot Springs

*Blowing Off Steam*

Geysers and hot springs occur when hot rocks deep inside the Earth cool down and create water vapor. The vapor is hot and it forces its way up until it finds a crack in the Earth's crust. It cools as it rises and eventually forms water which spurts out of a hole in the ground. Then it cools further and trickles back down the hole it came from. When the water hits the hot rocks at the bottom, it heats up and starts the whole process over again. This is why many geysers erupt at regular intervals.

*All Steamed Up*

The highest eruption was from the geyser called Waimangu in New Zealand. In 1903 it threw up steaming water 460 meters (1,500 ft) high, killing four people who had been standing as far as 27 meters (90 ft) away. Their bodies were strewn up to 800 meters (half a mile) from the geyser – one in a tree, one wedged between rocks, another jammed head first in the ground and the fourth on the ground. It seems that Waimangu was sorry for what it had done, as the geyser stopped erupting the following year and hasn't popped off since.

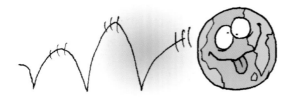

*Old Faithful*

This big old geyser in Yellowstone Park in the USA has been shooting a jet of water into the air once every 70–75 minutes for the last 80 years – hence its name.

# Trees and Forests

### Jungle Medicine

The world's largest rainforest is in the Amazon Basin in South America. The Amazon River runs through the area that covers over 7 million sq km (2.7 million sq miles). The rainforests and jungles of the Earth hold thousands upon thousands of species of animal, plant and insect. There is supposedly a cure for every ailment known to mankind within the plant life of the rainforest.

### Root Cause

There's a forest in Colorado, USA, that covers 0.4 sq km (106 acres) and in total weighs over 6,000 tonnes. Nothing strange about that, but for the fact that all the trees come from a single root.

Talking of roots – the longest tree roots are those of the South African wild fig tree which go down over 120 meters (400 ft) into the ground. You wouldn't pull that tree up in a hurry!

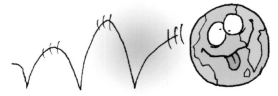

### Wood you Believe it?

Petrified wood isn't a log that's scared to death. It's actually wood that has been *fossilized*, which means that over the centuries the organic material (in this case wood) has been replaced by mineral deposits so it looks like, feels like, and *is* stone. There are whole areas of fossilized trees in some parts of the world and these are known as *petrified forests*.

### Old or Old?

In the Mojave desert in the USA, there is a plant which is 12,000 years old and still growing. But that's nothing – the oldest tree is thought to be a Tasmanian holly bush which has supposedly been growing for 40,000 years.

## Amazing Avalanches

Avalanches usually happen when a weak layer of snow gives way and tons of snow just slide right on down the mountainside. The highest number of avalanches, shifting millions of feet of snow, happen every year in the Himalayan Mountains, but these are rarely seen by humans because the area is so vast and unpopulated. Avalanches can move as fast as 80 km (50 miles) per hour across the ground.

## Deadly Fall

During the First World War, the sound of gunfire set off avalanches which killed between 40,000 and 80,000 soldiers in the Tyrolean Alps – more than were killed by bullets or disease.

## Drizabone

Lake Eyre is Australia's largest lake, but it's usually dry. To top it all, it's covered in a crust of salt, sometimes up to 4 meters (13 ft) thick, so it's not brilliant for watersports.

## Time to Dry Up

A lake in central Asia called the Aral Sea is likely to disappear in the twenty-first century. It was once the fourth largest lake in the world, but in the last 30 years, its water level have fallen by 14 meters (46 ft) due to water being taken for irrigation and industry from the rivers that feed it.

# Ancient H20

Water once covered the whole surface of the Earth. When the oceans were being formed 4,000 million years ago, it rained for over 60,000 years without stopping – now that *is* depressing! And, amazingly, that same water is still around today because it's being recycled all the time. The Sun dries the land and sucks the water up into the air, and it then falls as rain. And so it goes round, and round and round. In fact, most of the water on this planet is over 3,000 million years old, so if it tastes a bit funny, it's hardly surprising. Another incredible fact is that even with all the freshwater lakes, rivers and streams, only a tiny percentage (2.6 per cent) of the world's water is fresh, not salt.

## Storm Clouds

Floods can be caused by several factors – from glaciers melting to heavy rainfall. Clouds laden with water ready to fall as rain can weigh up to 500,000 tonnes and could be as high as 10 km (6 miles) above the ground. Flash floods can be the most dangerous because they usually build up high in the mountains and flow down dried-up river beds, so even if it's not raining where you are, you can still get caught by the flood.

The most people killed by a flood was in 1887 when the Huang He River in China burst its banks and killed 900,000.

27

## Too Much Stress

Floods can be caused by earthquakes in reservoirs, when tremors cause the water to spill over the dam. In fact, when engineers first started to build reservoirs, they noticed that earthquakes always happened just after the reservoir was artificially filled. They soon realized (but not soon enough for some people!) that the rocks surrounding the reservoir basin were being put under enormous stress from the pressure of the water. The way to get round this, it was discovered, was to fill the reservoir very slowly. That way, the rocks gradually became used to the pressure of water, and earth tremors were less likely to occur.

## Lumpy Bits

Another feature of reservoirs which can cause unexpected flooding is when huge lumps of rock fall from the surrounding high ground into the reservoir. The excess water pours down over the dam and flood the valley below.

## Power Shower

You might think that the idea of standing under a waterfall and having a natural shower sounds great, but if you were under the highest in the world – Angel Falls in Venezuela which drop 979 meters (3,213 ft) – the water would hit you at around 450 km (280 miles) per hour, enough to work up a right lather.

# WACKY NATURAL WORLD QUIZ

Did you enjoy that little collection? Take a minute to look at this little quiz. See how many of the questions you can answer and then try it out on your friends. I bet they won't believe that any of the answers can be right. You and I know different though, don't we!

1. **Earthquakes are caused by:**

a  Shifting crustal plates?

b  Volcanoes?

c  Everybody in China jumping up and down at the same time?

2. **A volcano that no longer erupts is known as:**

a  Dormant?

b  Extinct?

c  A bit sleepy?

**3.** The largest rainforest in the world is found in:

a   Africa?

b   South America?

c   Scotland?

**4.** Most of the water on the planet Earth is:

a   3,000 years old?

b   3,000 million years old?

c   Sold in bottles at the supermarket?

**5.** The Himalayan mountains are tall because;

a   They are young mountains?

b   They are very old mountains?

c   They ate lots of vegetables when they were young?

**6.** The highest waterfall in the world is:

a   The Victoria Falls?

b   The Angel Falls?

c   A great place to take a shower?

Answers on page 172.

# WACKY WORDS

## Scrabbling Around

Scrabble was invented in 1931 by Alfred Butts, an unemployed American architect. He called it Criss-Cross but he couldn't get anyone to buy the idea at first because the games manufacturers thought it was too boring. Butts and his partner made the first sets by hand in a garage and the game eventually went on sale in 1946. Scrabble has now sold over 100 million sets worldwide – not bad for a "boring" game.

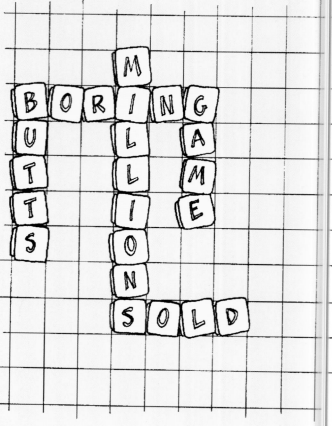

THE FULL MONTY SUITS YOU SIR

## The Full Meshe

Do you know where the phrase "the full Monty" comes from? It was coined way back when Montague Burton owned the men's clothes shop, Burtons. Sometimes people couldn't afford to buy a three-piece suit so they'd buy a two-piece (jacket and pants without the vest). A three-piece suit was something a bit special and became known as the full Monty. It's a bit strange then that it should end up being a euphemism for being completely naked!

The other strange thing is that Montague Burton wasn't his real name. It was Meshe Osinsky (he was Russian) and he changed it when he came to Britain.

# Famous Last Words

You know the feeling when you predict that something will or won't happen and the opposite occurs and you feel really stupid? Well these are a few whoppers that people just wish they hadn't said:

*"Heavier-than-air flying machines are impossible."*
Lord Kelvin, President of the Royal Society (in 1895).

*"Everything that can be invented, has been invented."*
Charles H. Duell, Head of the US Patents Office (in 1899).

*" Guitar music is on the way out."*
Head of Decca Records when they rejected the Beatles.

*"There is no reason anyone would want a computer in their home."*
Ken Olson, President of Digital Equipment Corporation.

# WACKY SUPERSTITIONS

## Abracadabra

If you want to bring good luck to something or ward off nasty spirits, then the way to do is to say abracadabra over and over, dropping a letter from the end each time you say it.

If you do the same thing and write it down, the words are in an inverted (upside down) pyramid shape, which is also supposed to be lucky.

Notice that the letters from the bottom A to the top-right A also spell ABRACADABRA.

```
A B R A C A D A B R A
A B R A C A D A B R
A B R A C A D A B
A B R A C A D A
A B R A C A D
A B R A C A
A B R A C
A B R A
A B R
A B
A
```

## Don't Burn Your Bridges

Never say goodbye to someone while you're standing on a bridge or you will never see that person again.

## Give Us a Twirl

If you've had a string of bad luck, then the way to change things for the better is to turn completely around in a clockwise direction three times (or seven times if things are really bad).

## Protective Face Paint

In some ancient cultures, especially those that worshipped a sun god, people drew a red circle round their mouths to make sure that their souls didn't fly out and to stop the devil from getting in, so the first lipstick was doing much more than just making your lips look kissable. Eyeshadow round the eye was also thought to protect people from the Evil Eye (although some eyeshadow makes women look like they've got a pair of 'em!).

## Inside Out and Back to Front

Another way to ward off bad luck is to turn your hat back to front (as horse-racing jockeys do) or your clothes inside out. (If you accidentally wear your T-shirt or sweater inside out you should have good luck.) You can also pull your pockets inside out to avert bad luck (also a sign that you've run out of cash). Spitting over your left shoulder and crossing your fingers helps to bring good luck.

## Good Luck

Anything made of gold brings good luck, as do rabbits' feet (not for the rabbit!) and birthstones (as long as you wear the right one). Bits of animal give you that animal's characteristics. For example, you should carry a lion's tooth for courage and a fox's tail for speed.

WHERE'S THE HANDLE-BARS?

## Getting Stoned

Did you know that the semi-precious stone *amethyst* (the purple one) was thought by the Greeks to prevent drunkenness if you wore one? The word *amethystos* means "not to be drunk."

In the Middle Ages it was known as the Bishop's Stone. It's the birthstone of Aquarians and is supposed to make them more self-disciplined.

Crystals also bring good luck and protection; jade protects and is good for fertility. It is also placed on the eyelids of a dead person to help their soul to find its way in the afterlife.

## Chris-Cross

If you thought it was ungodly or wrong to shorten the word "Christmas" to "Xmas" then don't worry, there's nothing wrong with it after all. The X is a sacred symbol and so it's okay to use it instead of Christ.

## On the Cards

If a gambler is having a bad run in a game of cards for example, then a few turns in a clockwise direction should put him on to a lucky streak. The other thing for gamblers to remember is never play with a cross-eyed partner or you'll lose.

For luck, you can rub your dice on the head of a person with red hair. Red hair is generally thought to be lucky and will bring luck to anyone who strokes it (although many ancient cultures believed redheads to be witches). If you borrow money to play cards you can't lose, but the reverse is true if you lend money. You'll also lose if you drop a card while you're playing.

RUB! RUB!

WHAT A LOVELY BOY

## Pretty Stupid

In some cultures people don't like anyone to praise their children because they think that if the devil overhears, he'll come and get them. But if you're not too clever and not much to look at you're all right – the devil isn't going to want to steal you away from your parents. Jewish people immediately say a protective phrase which means "may no harm come to you," to counteract a compliment given to their child. The Chinese go so far as to say the opposite is true if their child is praised, as in "Oh no, she's really very ugly and very,very stupid." So if you're Chinese and you think your parents hate you, they're only trying to help.

## Ear Ear

The ear is supposedly the center of intelligence – probably because it leads through to the brain! Pulling children's ears to make them remember their lessons probably comes from this ancient belief. Sailors thought that wearing an earring would help their eyesight and protect them from drowning. Sorry, sailors, that's a *life* ring you're getting confused with there, and it's a bit big to wear in your ear.

## Flagging Troops

Did you know that superstition says that flags should never touch the ground? During battles in days of yore, the flag bearer was the chap who had to keep it up all through the fight. If the flag was lowered it meant that the king had been killed or injured. (It didn't, of course, mean that the flag bearer had snuffed it and therefore couldn't hold the pole any longer!)

## Catch a Druid by his Toe

Eeny Meeny Miny Mo was thought to be an ancient magical rhyme used to choose human sacrifices for Druidic ceremonies. Not such an innocent sounding nursery rhyme after all.

## OK Coral

If you want to stop the rows at home stick a piece of coral on the mantelpiece – it's supposed to bring about peace and tranquility and to prevent nightmares if you keep it by your bed.

## Night Air

Night air was once thought to be unhealthy. It was believed that all sorts of poisons seeped out of the ground after the sun had set.

# Well Broomed?

Whatever you do, don't leave any unwanted bits of your body lying around. If a witch gets hold of your nail cuttings or eyelashes or trimmed bits of hair, she can cast a spell on you. On the other hand, you can use your fallen eyelashes to make nice wishes for yourself like this:

Put the eyelash on the back of your left hand and make the wish. Then put your right palm onto the lash and press down. Your wish will definitely come true if your lash sticks to the palm of your right hand.

## Time for a Manicure

You can't cut your nails on just any old day – it's got to be on the right day of the week. Monday is a good day, Friday and Sunday, particularly Sunday, not so good. And don't for goodness sake cut them in any kind of order – that causes all sorts of nasty things to happen.

## Leave it Behind

Did you know that it's bad luck to go back for something you've forgotten? So next time you forget your homework, you can tell the teacher that you would have gone back for it but . . .

## TGI not Friday

Fridays are meant to be unlucky (so why do so many people say "thank God it's Friday"?!). Thirteen is also very unlucky, so the combination of Friday the thirteenth is not a good one.

## Lucky Seven

Horseshoes are thought to be lucky because they are arch-shaped (like many church windows and doors) and also because they are normally hammered in with seven nails and seven is the luckiest number.

DINK!

## Uncool Cat

In 1999 a cat was baked in an oven at 66 degrees centigrade (151 °F). The cat singed his paws badly but lived to tell the tale (tail?). He was given the nickname Scorcher; the vet who looked after him was called Mr Cook!

Cats have always played a big part in mystery and superstition, and have often been linked with witchcraft and black magic (through no fault of their own). For example, it's either lucky or unlucky if a black cat crosses the road in front of you, depending on who you've been listening to.

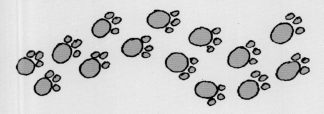

## Cat-ching Up

Sam the American cat didn't like being left behind when his owners moved from Arizona to Wisconsin, nearly 2,400 km (1,500 miles) away. It took Sam four years and a lot of paw-work before he was reunited with his family.

## In Mourning

What would you do if you were an ancient Egyptian and your cat died? Shave your eyebrows of course. It was against the law not to do so and you had to mourn for a period of time as well (which is what shaving the eyebrows was all about). If you killed a cat, even if it was an accident, the punishment was death.

WISCONSIN 1,000M

## Dog Days

Dog Days were the hottest, most fly-ridden, unhealthy days of the year according to the Romans. They are called the "days of the dog" because they occur when the star Sirius, also known as the Dog Star, is in a position where it adds its own heat to the heat from the Sun, making the weather very hot indeed. Dog Days are from 3 July to 11 August.

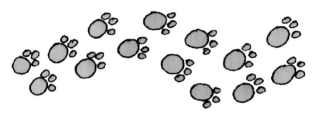

## It's a Dog's Life

Gunther the German Shepherd dog inherited a whacking $65 million (£40 million) from his deceased owner, a German countess. Gunther, who lives in Italy and is chauffeur-driven in his BMW sports car, is the richest dog in the world.

## Frog Snog

Some people think frogs are lucky, that they bring friends and are a symbol of fertility (probably because they lay huge numbers of eggs!). In some cultures, frogs hold the souls of dead children and are not to be trusted, and in others a frog will turn into a handsome prince if you kiss it (ugh!).

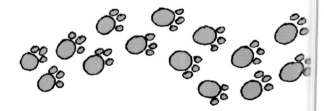

## Feeling Blue

Don't touch that little blue frog though – he's one of the most poisonous animals in the world, hence his name: the poison-arrow frog. He might look very pretty, but the poison on his skin is absolutely lethal. The *golden* poison-arrow frog is even more beautiful – and even more toxic.

## Toad in the Hole

There are stories of frogs and toads that can apparently survive for a very long time trapped inside rocks. When the rocks have been split open by stonemasons or miners, the animal happily jumps out. One theory is that the frog or toad climbs in through a small hole in the stone when it is still tiny, then grows a bit so it can't get out again. One toad had been entombed in the cornerstone of a courthouse for more than thirty years. Another was found alive in a 2-tonne block of limestone but died as soon as it was released (probably from shock!).

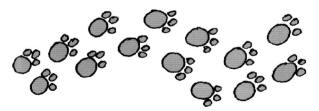

## White Lie

Did you know that polar bears are not really white? They're actually colorless and each hair is hollow. Air is trapped in the hollow tubes and helps to keep them warm.

## Toxic Toads

As a defence against predators, the Australian cane toad squirts a hallucinogenic liquid (which makes you think that you're seeing things that really aren't there). Some sad people have been known to catch the toads so they can have a lick and get a kick from the toad juice.

## Big Baby

A baby blue whale suckles 460 liters (1,012 pints) of milk per day from its mother (remember that whales are mammals like us, only we're a bit smaller and don't spend so much time in the water).

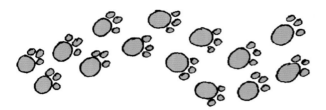

## Gone But Not Forgotten

The last *quagga* died in August 1883. What was it? A relation to the zebra which was first spotted (or should we say, *striped*? ha, ha) in 1685. She died in Amsterdam Zoo and no one realized that she was the end of the line.

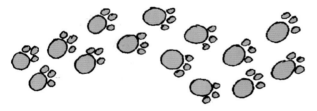

## If Looks Could Kill

In the old days, it was thought that some people had the power to kill just by looking. In some cultures it is still believed that a snake can do just that, because it can poison the air with its eyes (scary, but not very believable).

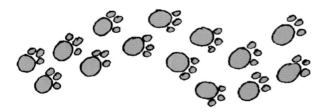

## Gift Horse

The saying "never look a gift horse in the mouth" doesn't mean you should never turn down something that's being offered to you. What it does mean is that you shouldn't inspect it too closely. This comes from the practice of telling the age of a horse by checking its teeth. If it's ancient it's not really worth having after all.

# Macho Tortoise

Perhaps it's time to give the poor old tortoise a different image. An adult giant tortoise shell can weigh at least 200 kilos (440 lb) which they shoulder with no fuss at all, whereas we humans would consider a backpack weighing 27 kilos (60 lb) a heavy burden. And they have to carry their shell around for up to 150 years, so it's not surprising that they move so slowly.

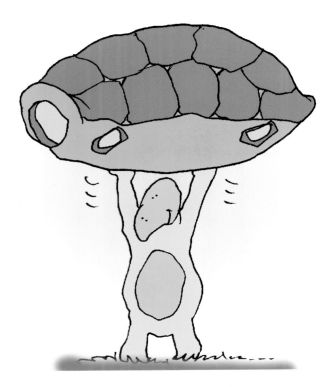

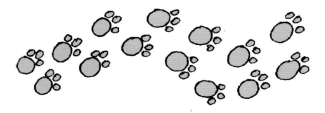

# Gotta Lotta Bottle

Bernard Keyter of the South African Institute for Medical Research milked 780,000 poisonous snakes over a period of 14 years and got 3,960 liters (870 gallons) of venom (poison) out of them. He milked each snake himself, by hand, and was never bitten.

# Mistaken Marsupial

When Captain Cook got to Australia, the first thing he did was to ask an Aborigine what on earth were those strange creatures hopping all over the place with their babies in their coat pockets. The Aborigine replied "kan-goo-roo", which is probably not the name of the animal but the Aborigine saying, "I don't know what you're on about, buddy."

## Dirty Rat

Did you know? The rat is the most dangerous small mammal because of all the diseases it carries. The Black Death (bubonic plague) was carried by rats and in the course of 300 years it killed 25 million people in Europe alone. The brown rat, which is known as the Norway rat, and the black rat or ship's rat (whose Latin name, believe it or not, is *Rattus rattus*!) are both guilty parties, but it is the flea that lives on them that is responsible for carrying the bacteria.

## Fast Birds

An ostrich can run at up to 72 km (45 miles) per hour, which is faster than the speed of the average car journey. Instead of eating them we should be riding them! If you think that's fast, how about this: a flying swan could overtake a cheetah running at full pelt (excuse the pun!).

## Lost or Slow?

Blue Clip, the homing pigeon, took seven years and two months to fly home from France to Manchester, England, a distance of 595 km (370 miles).

# Is it a Bird, is it a Plane . . .?

In 1886, hunters in South America killed a huge unknown bird and measured its wingspan. Six men stood side by side with their arms outstretched and only just made the distance from wingtip to wingtip. It was about 10 meters (33 ft) – not the sort of bird you'd keep in a cage in the front room, but could this have been the mythical *thunderbird* that they killed? If so, was it the last one in existence?

## Bird Luck

There are many superstitions based around animals and birds. It is said that if a bird flies into the house you'll hear important news shortly afterwards, but it's bad luck to have wallpaper with birds on it in your home – not to mention bad taste. White birds are supposedly a sign of death, so don't go down to the seashore because there are loads of 'em there. And the white birds following behind the farmer's plow are surely a sign of death – for all the worms and insects in the field!

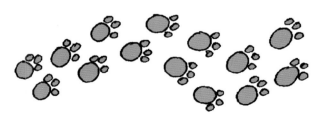

## Tell Tale Tit

A Blue Jay supposedly spends every Friday with the devil, telling him all the bad things we've done in the week (surprised he's got any time to eat or to do anything else, then).

# WACKY WORLD OF CREEPY CRAWLIES

### False ID
A spider is not an insect because it has eight legs not six – it is an arachnid, as are scorpions, mites, ticks and so on.

## Cricket Luck

Many cultures across the globe believe in the luck of the cricket (grasshopper). It's very bad news to kill a cricket because it is believed to be a house spirit and will bring good luck. And crickets don't like being copied, so next time you feel like sitting in the grass, rubbing your legs together to make a funny noise, don't – or the crickets will send the boys round.

## Snail's Pace

In the time it would take a snail to cover 100 meters (328 ft), a sprint athlete could do the same distance 700 times over, because the fastest any snail can go is only 50.3 meters (165 ft) per hour. Mind you, he *is* carrying his whole house around on his back.

## Bee-Line

If a bee flies into your house it's good luck, but don't let it stay there and die or there'll be trouble. If a whole swarm of bees turns up on your doorstep it's very bad news (yes, we'd agree with that!).

## Aged Arachnid

Did you know? Female tarantulas can live as long as 30 years. However, it's difficult for scientists to find out exactly how old they are because they always lie about their age (they all say that they're 21!).

## Extra Long

Talking of worms, do you know that a common or garden earthworm pulled out of the ground by a l'il ole blackbird in Britain in the 1980s was 2 meters (6.6 ft) long?

## Charming Worms

Would you believe that there is actually a World Worm Charming Championship? It's held at Willaston, Cheshire, UK. In 1997, Tom Shufflebotham charmed 511 worms out of the ground in just 30 minutes. Surely easier just to open a can of them?!

## The "Bee-Stung" Look

Frank Bornhofer was photographed wearing a "helmet" of bees, complete with bee chinstrap. He did it to prove that bees rarely sting, which is a drastic measure and who cares anyway, but luckily for him, they didn't!

## Sting in the Tail

The biggest scorpion lives in West Africa. *Pandinus imperator* is as big as a man's hand and weighs up to 60 grams (2 oz). But the scorpion with the most poison is the fat-tailed scorpion of Tunisia. So be warned. If a fat-tailed scorpion ever asks you "Does my butt look big in this?", say NO.

## Food for Thought

It isn't just an old wives' tale that says fish is good for your brain and the more fish you eat the smarter you will be. Like fish, the human brain contains a fair amount of the chemical phosphorus. So applying the principle "you are what you eat," if you consume lots of fish, you should become a nuclear scientist or a similar clever Trevor. Either that, or you'll end up with two googly eyes, a slimy skin and enjoy spending most of your time in water.

## Relatively Large

It's a bit worrying that a lot of the insects we have in our gardens are actually related to the fish and crustaceans which live in the sea. What's going on?! For example, the giant squid is actually a relative of the garden snail. (That's fine, as long as Big Squid doesn't come to visit his long-lost cousin Brian in *my* garden!)

## Fishy Beliefs

In some countries people believe that the fish is sacred and a symbol of fertility; in others it is considered food for ghosts, and in some cultures it is forbidden to eat fish at all. However, if you do eat fish, eat it from head to tail for good luck.

**Teething Pains**
Piranha fish have large jaws, very sharp teeth and they like the taste of raw flesh, especially human. In Brazil in 1981, more than 300 people were eaten alive by these vicious little meat eaters when their boat capsized in the dock.

**Father Figure**
The male seahorse carries its babies around in a bloated pouch in its front, very much like a pregnant woman. The female seahorse goes out to work and wonders why the seahorse stable is in such a mess when she gets home. (Yeah, yeah.)

LOOK AT THIS PLACE!

KIDS TODAY!

**Still Around**
The *coelacanth* is probably the oldest fish around. It lived about 60 to 70 million years ago (around the time the dinosaurs disappeared), and it was thought to be extinct, but now appears to be alive and swimming in the Indian Ocean.

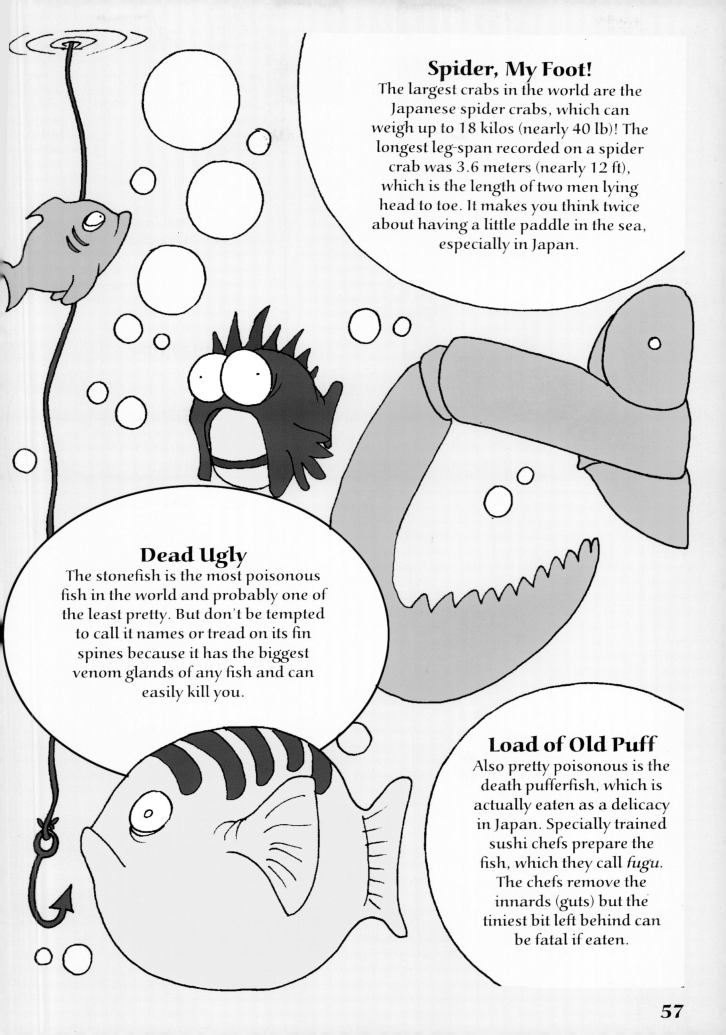

## Spider, My Foot!

The largest crabs in the world are the Japanese spider crabs, which can weigh up to 18 kilos (nearly 40 lb)! The longest leg-span recorded on a spider crab was 3.6 meters (nearly 12 ft), which is the length of two men lying head to toe. It makes you think twice about having a little paddle in the sea, especially in Japan.

## Dead Ugly

The stonefish is the most poisonous fish in the world and probably one of the least pretty. But don't be tempted to call it names or tread on its fin spines because it has the biggest venom glands of any fish and can easily kill you.

## Load of Old Puff

Also pretty poisonous is the death pufferfish, which is actually eaten as a delicacy in Japan. Specially trained sushi chefs prepare the fish, which they call *fugu*. The chefs remove the innards (guts) but the tiniest bit left behind can be fatal if eaten.

# WACKY WORLD OF ANIMALS
## QUIZ

Did you enjoy that little collection? Take a minute to look at this little quiz. See how many of the questions you can answer and then try it out on your friends. I bet they won't believe that any of the answers can be right. You and I know different though, don't we!

1. **When a cat died in ancient Egypt its owner would:**

a  Bury it?

b  Shave their eyebrows?

c  Turn it into a scarf?

2. **Dog days are:**

a  The hottest days of the year?

b  The coldest days of the year?

c  When dogs go on their holidays?

**3. A quagga was:**

a  An animal related to giraffes?

b  An animal related to zebras?

c  An animal related to your uncle?

**4. The world's most poisonous scorpion is known as the:**

a  Fish-tailed scorpion?

b  Fat-tailed scorpion?

c  Run-away-as-fast-as-you-can scorpion?

**5. The giant squid is related to:**

a  The garden snail?

b  The earthworm?

c  Grandfather giant squid?

**6. The stonefish is the:**

a  Most poisonous fish in the world?

b  Most beautiful fish in the world?

c  Hardest fish in the world?

Answers on page 172.

**59**

WACKY

PLANTS

## Popping Up

The soldiers who dug the trenches during the First World War created fields of poppies, because poppy seeds can lie dormant for years, only springing into action when the earth is disturbed. That's why Remembrance Sunday is recognized with artificial poppies for sale.

## Racing Green

Bamboo is a type of grass and can grow nearly 1 meter (3.3 ft) per day! In some countries, bamboo was used as a form of torture. The prisoner was tied flat to the ground and the bamboo eventually grew through his body while he was still alive.

## Weather Forecaster

Dandelions have something in them that makes you pee (the posh word for it is *diuretic*) and were known during the Middle Ages as "piss a beds."

You can also use dandelions to forecast the weather, because if you see them flowering in April *and* July, it'll be a hot and steamy summer. And you can tell the time by blowing the fluffy seed heads, and infuriate your dad by spreading seeds all over the garden.

ONE
O'CLOCK

TWO
O'CLOCK

## Weedy Drink

Dandelion and burdock is a great drink (they love it "up North" in Britain) and so is dandelion coffee. You can drink it all day without getting the trembles, because there's no caffeine in it.

# Number One Seed

The seed of the sea palm found in the Seychelles is the largest in the world and can weigh as much as 18 kilos (40 lb)!

# Heavy Veg

How about this for some wacky vegetables? The longest recorded carrot was nearly 5.5 meters (17 ft) long and the heaviest celery weighed 21.8 kilos (48 lb 1 oz).

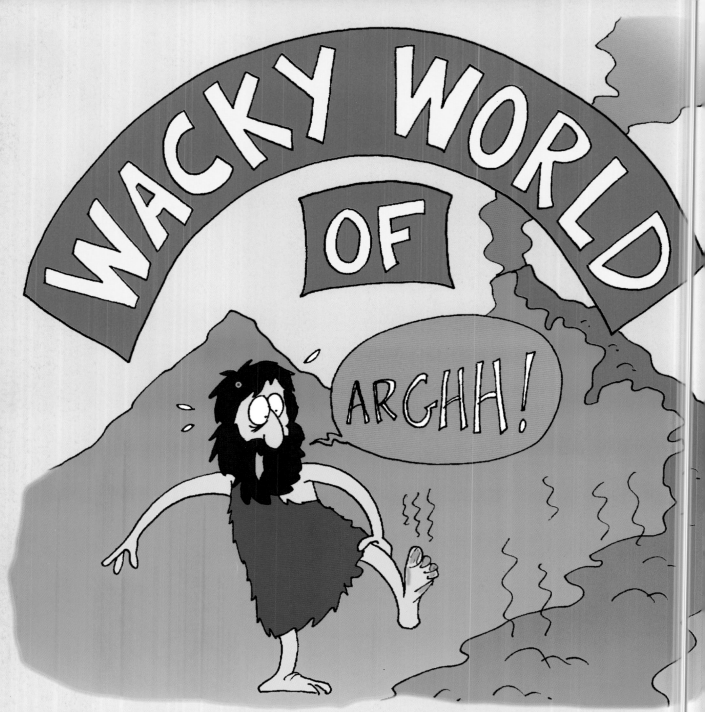

## Ancient Footprints

It's pretty amazing when you think that although "civilization" has only been happening for a few thousand years, man was around long, long before that. The anthropologist Mary Leakey found the oldest humanoid (like a human, rather than an ape) footprint in Tanzania in 1978. It was preserved in volcanic lava and had been made 3.7 million years ago!

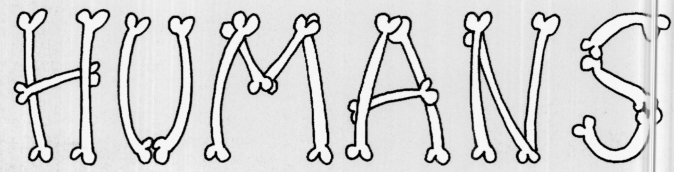

# Wacky Nuts

## Child's Play

Lord Lytton was a little soft in the head. He really believed that he could make himself invisible by covering his face with his cloak, just like a little child. When he had friends to stay in his house, he'd walk around with his cloak over him and his friends would pretend that he wasn't there. When he threw his cloak off, he would expect everyone to be surprised that he'd "suddenly" appeared.

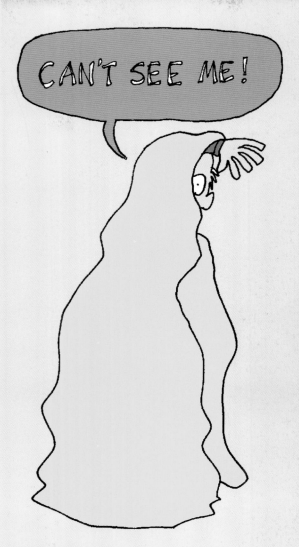

## Barking Mad

Ludwig II of Bavaria had funny ideas about night-time and daytime, and would get up at night to start his day! His brother, Otto, was equally bonkers; he believed that if he didn't shoot a peasant a day he would become ill. He spent the rest of his time howling and barking like a dog.

## No Room to Move

A Californian woman believed that she had to keep adding rooms to her house because if she didn't she would die. She spent $5 (£3) million over 38 years and created 160 rooms in her house, some only a few inches wide. It's probably one of the world's oddest buildings, with 10,000 windows, 2,000 doors and 9 kitchens.

## Shocking Behaviour

When electric chairs were first invented, Menelek II of Abyssinia ordered three and then remembered that there was no electricity in his country, so he used one of the chairs as his throne. Menelek was also a bit of a paper muncher and died after eating a particularly thick Bible.

*Red Flag*

There were only two car drivers in Britain in 1894. The first of these was Henry Hewetson, who bought his two-horsepower (not very fast) car for $120 (£80). Henry drove around his native Catford for six weeks until the police decided that he could only drive if he had a man walking ahead of him with a red flag. To get round this, he hired a man with a bicycle to ride ahead of him to look out for policemen, and another to sit in his car with a flag ready to jump out if a cop was spotted.

## Wacky Mistakes

In 1982, Michael Scaglione missed his putt at the thirteenth hole while playing a round of golf. He was so furious that he hurled his golf club at the golf cart in a temper. It broke in half against the cart and rebounded, and he bled to death from a stab wound in the neck. The moral of the story is either: 13 *is* unlucky for some *or* don't throw a tantrum when you lose!

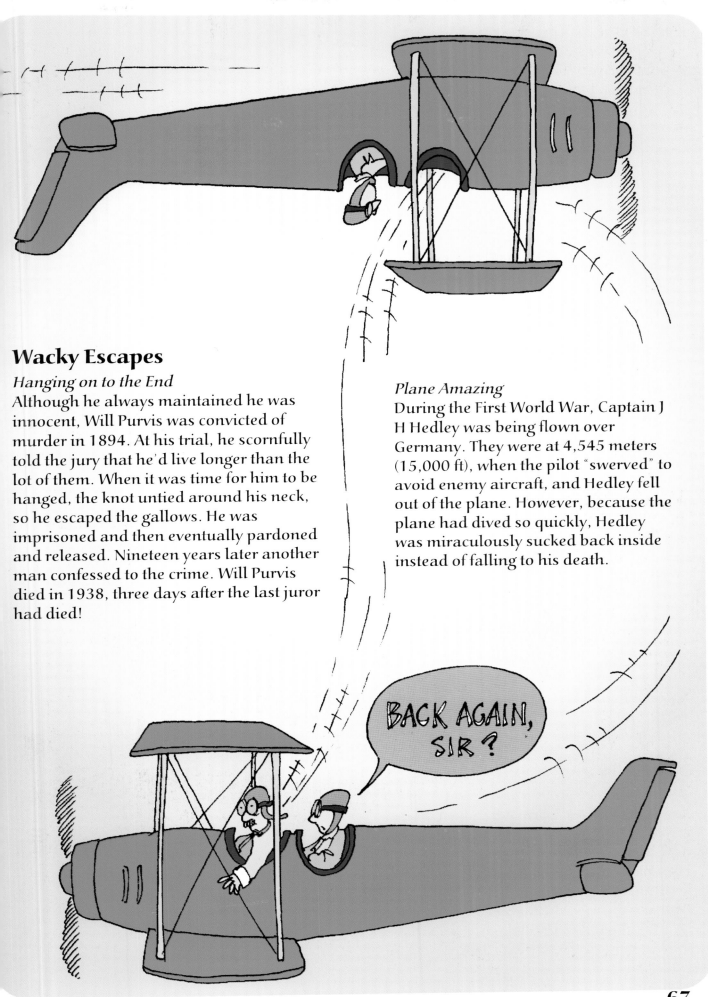

## Wacky Escapes

*Hanging on to the End*

Although he always maintained he was innocent, Will Purvis was convicted of murder in 1894. At his trial, he scornfully told the jury that he'd live longer than the lot of them. When it was time for him to be hanged, the knot untied around his neck, so he escaped the gallows. He was imprisoned and then eventually pardoned and released. Nineteen years later another man confessed to the crime. Will Purvis died in 1938, three days after the last juror had died!

*Plane Amazing*

During the First World War, Captain J H Hedley was being flown over Germany. They were at 4,545 meters (15,000 ft), when the pilot "swerved" to avoid enemy aircraft, and Hedley fell out of the plane. However, because the plane had dived so quickly, Hedley was miraculously sucked back inside instead of falling to his death.

BACK AGAIN, SIR?

## Third Time Lucky

Robert Clive was only 19 when he tried to commit suicide, but he failed twice because his pistol wouldn't work. He went on to become the famous Clive of India, but in 1774 – 30 years later – he tried suicide again – this time successfully.

## Wacky Healing

*Eyeball*

Did you know that some people can tell what's wrong with a person by looking into their eyes? The science is called *iridology* and it works by looking at the shape and condition of the iris (a part of your eye, not the flower!).

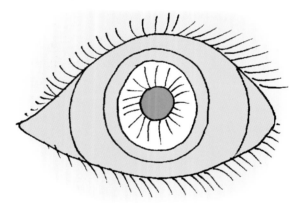

*No Tools*

Some people who call themselves *psychic surgeons* claim to be able to perform operations without the use of anesthetic or surgical instruments! These psychic surgeons often don't need to touch their patient, but perform the operation in the air above the affected part. When instruments are used, often the wounds heal *very* quickly afterwards, and the patient becomes well again. However, there are a few quacks who are in it for the money and who use trickery, such as sneaking in some animal blood and tissue to pretend that they've opened the patient's body and extracted the unwanted bits.

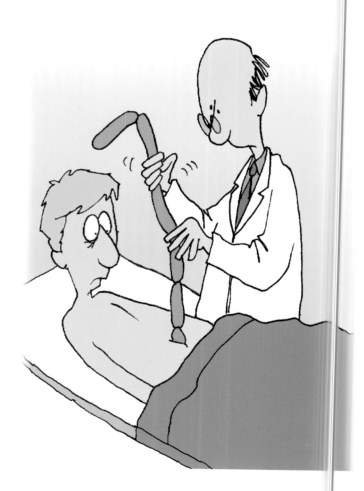

## Psychic Sidekick

A psychic surgeon who performed a lot of operations between 1950 and 1971 was a Brazilian, Jose Arigo, who said that when he was performing operations his mind and body were taken over by a long-dead German doctor. He never used anesthetic, but his patients felt no pain and were conscious while he cut them open. In fact, he would often remove some nasty unwanted bit from the patient and give it to them to look at. He didn't sterilize his instruments but his patients never had any infection from the operation, and their wounds healed more quickly than was normal. Arigo became a bit of a hero and his miracle healings were reported in the national newspapers.

## Wacky Money

*But I Thought it Was Mine!*
An American roofer found an extra $88 million (£58.6 million) in his bank account. Sadly for him it was a mistake, but that didn't stop him initially drawing out $4 million (£2.7 million) before he gave the whole lot back.

### A Bob or Two

The Sultan of Brunei is worth $30 (£19) *billion* and Bill Gates, the Microsoft man, is almost as rich. The Queen of England is supposedly the richest woman in the world but she's only got $400 (£250) million. However, top prize goes to the millionaire who has given the most money away. Viscount Nuffield, the car manufacturer, donated $48 (£30) million to various causes.

### Gold Weight

The expression "worth his weight in gold" came from a birthday tradition in the East. For example, the Aga Khan III on his sixtieth birthday topped the scales at 110 kilos (243 lb), and received that amount in gold and diamonds. It only works if you or the person giving you the present is rich. For most of us it would more likely be "He's worth his weight in pennies."

### Mean Time

How's this for stinginess? When she died, the American millionairess Hetty Green was worth what today would be well over a billion dollars (£816 million), but she still tried to find a *free* hospital which would operate on her son. She didn't find one in time so he had to have his leg amputated. She herself ate her food cold so that she didn't have to pay for the cost of heating it.

## Look At My Wad!

Walter Cavanagh of California, USA, has 1,397 credit cards. They're worth more than $1.65 million (£1 million) in credit and are kept in the world's longest and heaviest wallet. It's 76.2 meters (250 ft) long and weighs 17.49 kilos (38 lb 8 oz).

# Wacky Stunts

## Eau No You Don't

Evidence suggests that a French soldier imprisoned on a ship in Dover, England, escaped by swimming to France in 1815. But no one knows for sure if it was true, because he wasn't going to own up to it.

## Water Performance

Since the French prisoner who might have been the first man to swim the English Channel, there have been plenty of people (nutters) who have slathered themselves in sheep fat and waded in for a minimum 24-hour swim. But that's nothing compared with the longest swim on record. An American, Fred Newton, was in the water for 742 hours in order to swim 2,938 km (1,826 miles). He must have been as wrinkled as a prune when he climbed out of the river.

## Unique Talent

In 1952, at the Bertram Mills Circus, Rudy Horn, a German unicyclist, was able to balance on his unicycle while throwing six cups and saucers onto his head with his feet. He balanced them there while he added a teaspoon and a lump of sugar. How on earth did he do that?

## Scaling New Heights

The highest a man has ever reached is 397,848 km (248,655 miles), which is 253 km (158 miles) higher than the Moon. The fastest speed at which humans have ever travelled is 39,897 km (24,791 miles) per hour (in space, not on land!).

## Balancing Act

Amresh Kumar Jha of Bihar, India, balanced on one foot (his own!) for 71 hours and 40 minutes. That's 20 minutes less than three whole days and nights.

## Free Ride

German hitchhiker Stephan Schlei travelled 807,500 km (over half a million miles) for free in a 25-year period.

I NEED TO GO TO THE LOO!

WE'LL HAVE TO GO BACK, I FORGOT MY BRIEFCASE

## Taxi!

The longest cab ride was 34,908 km (21,69 miles) from London, UK, to Cape Town, South Africa, and back. It only cost $64,19 (£40,210).

## Long Bike

The longest motorcycle in the world is 7.6 meters (24 ft 11 in) long, built by Doug and Roger Bell of Perth, Australia. It only has a 250 cc engine, unlike the longest British motorbike, which is *only* 3.81 meters (12 f 6 in) long but has a V8 engine with a cubic capacity of 3,500!

## Get Off!

In 1910, Rama Murti Naidu allowed a fully grown elephant to stand on him, just to prove he could take it. The elephant weighed 3,176 kilos (3.1 tonnes) and it stood on a plank of wood which was laid across Naidu's body.

## Combi-dextrous

Thea Alba was a German schoolgirl who could write different words at the same time with her hands, feet and mouth. Very handy (no pun intended!) if you've left your essay until the last minute and need to write one in treble-quick time.

73

# Wacky Punishments

*Will They – Won't They?*

A Japanese murderer, convicted of poisoning 12 people, spent 39 years on Japan's death row in Sendai Prison. He died at the age of 94, probably from the suspense of it all.

## Punishing Punishments

The Babylonian king, Hammurabi, devised a code of law in 1750 BC. The punishments for breaking the laws were pretty stiff. For example, if a man killed a pregnant woman, his own child would be killed. That sounds fair (not!). And if a builder built a house so badly that it fell down and killed any of the inhabitants, the builder was put to death. That one could be useful nowadays.

## Boiling Point

The Bishop of Rochester's cook was boiled to death in 1530 for poisoning two members of the Bishop's household. He would have known only too well how those lobsters felt when he cooked them for the Bishop's supper.

# WACKY WORLD OF HUMANS QUIZ

Did you enjoy that little collection? Take a minute to look at this little quiz. See how many of the questions you can answer and then try it out on your friends. I bet they won't believe that any of the answers can be right. You and I know different though, don't we!

1. Menelek II of Abyssinia died after eating:

a  A poisonous fish?

b  A Bible?

c  So much that he exploded!?

IRIDOLOGY DEPARTMENT

2. The science of iridology is concerned with:

a  The eyes?

b  The ears?

c  Getting rid of things?

**3. The richest woman in the world is thought to be:**

a  Madonna

b  Queen Elizabeth II of Britain.

c  Married to the richest man in the world?

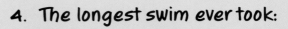

**4. The longest swim ever took:**

a  742 hours?

b  742 days?

c  742 years?

**5. The longest motorbike in the world is?**

a  76 meters long?

b  7.6 meters long?

c  Very difficult to park?

I'M DONE!

**6. The Bishop of Rochester's cook was boiled to death for:**

a  Poisoning the King?

b  Poisoning two of the Bishop's staff?

c  40 minutes at gas mark 7?

Answers on page 172.

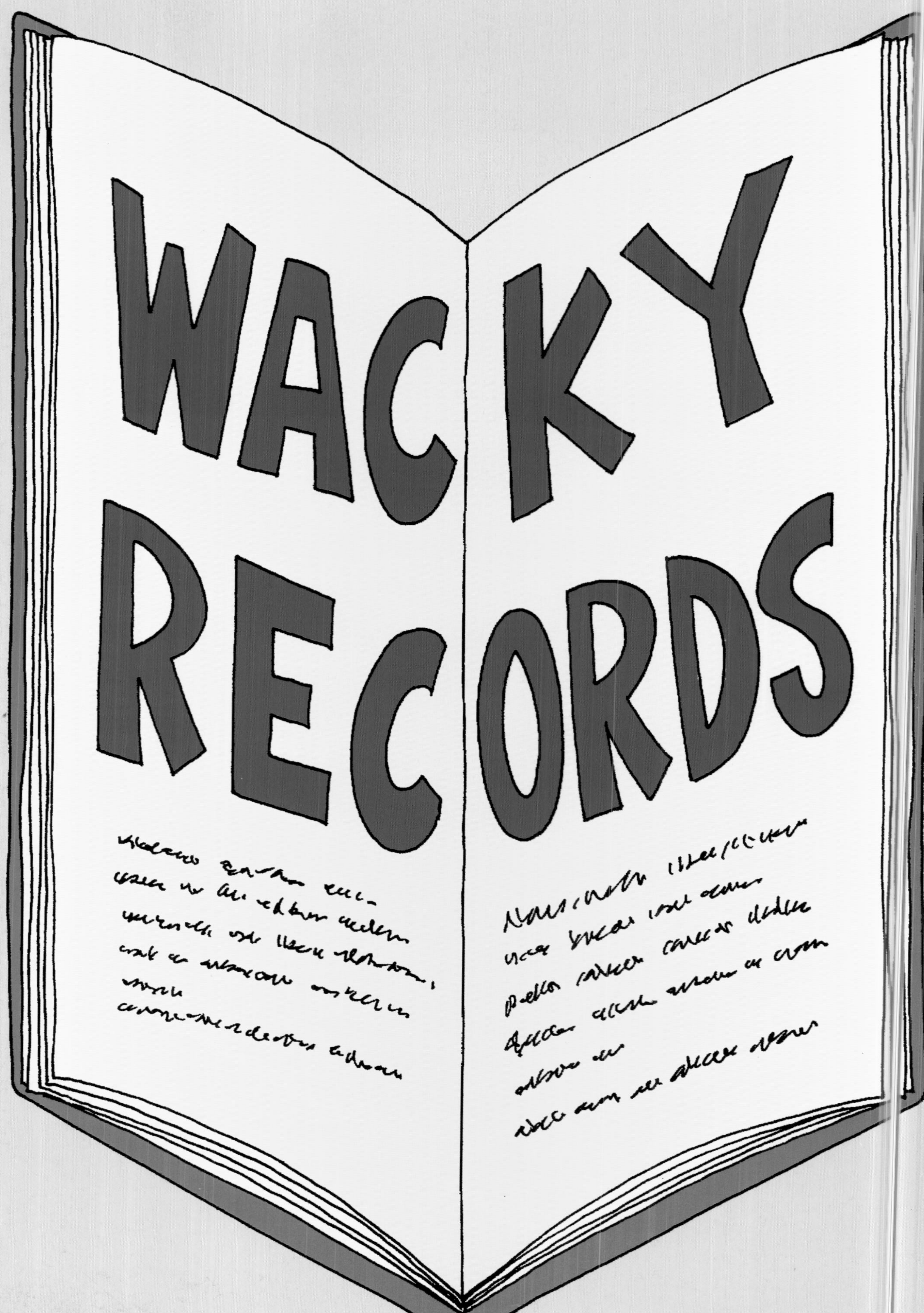
WACKY
RECORDS

## Hop On!

The first escalator at Earls Court subway station in London was used in October 1911. A chap called Bumper Harris, who had a wooden leg, was paid to go up and down the escalator to show how simple it was to use, even if you did only have one leg.

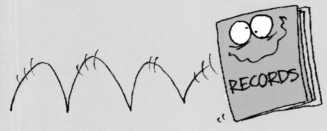

## Aptly Named

The first pilot to be killed in warfare was a Bulgarian who was shot down on 3 November 1912. His name was M. Popoff.

I SEE POPOFF'S POPPED OFF

## That's Gratitude

The first policeman to be murdered on duty was Metropolitan Police Constable William Grantham. He saw two drunk Irishmen fighting over a woman; he tried to help, but was beaten to death by all three, including the woman.

## Ancient Lights

Minnie Munro was 102 when she married her young man. Dudley Reid was only 83. What a cradle snatcher! Two other oldies who thought that life begins at 90 were Simon and Ida Stern, who were 97 and 91 years old respectively. They decided to get divorced, not having quite enough patience to wait "until death do us part".

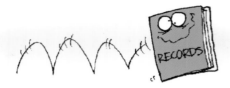

## Long Stretch

Polly Gadsby started work at the age of nine. Eighty-six years later, at the age of 95, she was still wrapping elastic.

## Pure Genius

Ganesh Sittampalam, the youngest graduate in Britain this century, got his math degree at the age of 13 years, 5 months. Another clever chap was Matthew Trout, from the UK, who began an Open University mathematics course at the age of 10 years and 10 months. And Zerah Colburn, a six-year-old whizz-brain from Vermont, USA, could multiply 12,225 by 1,223 (and come up with the right answer!).

## Leader of the Pack

Dominic O'Brien from the UK could memorize a single pack of shuffled cards in 38.29 seconds after looking at them just once. But American Dave Farrow memorized 52 packs of cards all shuffled together. There were 2,704 cards to remember and he only got six wrong.

## Facial Hair

Janice Deveree, the "Bearded Lady" of Kentucky, USA, had a 36-cm (14-inch) beard measured in 1884. Presumably she was proud of it, or she wouldn't have admitted to it being so long.

## Baby Boom

The most children born to one woman is 69. The Russian peasant woman was pregnant *only* 27 times but she had lots of twins and triplets and quadruplets. The highest number of recorded pregnancies for a woman is 38, resulting in the birth of 39 babies!

## You've Got to Hand it to Him

Shridhar Chillal last cut his fingernails in 1952, so when they were measured 45 years later, the longest one (the thumb) was 1.4 meters (4 ft 4 in) in length. The rest of the fingernails were almost as long and the combined measurement was 6.12 meters (20 ft 1 in)!

## Tall Story

The world's tallest woman was Chinese, believe it or not! By the time Zeng Jinlian was four years old she was over 1.56 meters (5 ft) tall, and when she died in 1982 she was just under 2.48 meters (8 ft 2 in) tall.

## Suits You, Sir

A three-piece suit was made from scratch (well, wool actually) by 65 members of the Melbourne College of Textiles, Australia. They caught the sheep, sheared it, washed and wove the wool, then cut the cloth to a pattern and stitched it in just 1 hour, 34 minutes and 33.42 seconds.

## Des Res

The longest car in the world is a 30.5-meter (100-ft) 26-wheeled limousine. It was designed by a Californian man, Jay Ohrberg, and has a swimming pool with diving board and a king-sized water bed. So if you fancy a splash while making a dash this is the car for you.

# Baby Weights

The lightest surviving baby was born in 1987 and was under 280 grams (10 oz), less than some bags of sweets! The heaviest, born in 1955, was 10.2 kilos (22 lb 8 oz). It's enough to make your eyes water!
*While on the subject of babies, here are some superstitions surrounding them:*

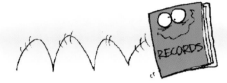

To make sure you have a girl baby, put a frying pan under your mattress; for a boy, place a knife there.

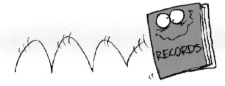

A pregnant woman should always eat the food she craves or the baby will be born with a birthmark in the shape of that food.

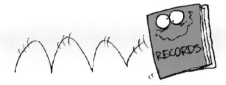

Don't let the baby look in a mirror before it is one year old (or it'll see how ugly it is!). Don't cut the baby's fingernails before it is a year old (the mother must bite them off instead) or the baby will turn out to be a thief.

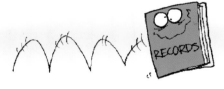

Mothers-to-be should not witness ugly events, because children are marked by what their mothers see, hear and do during pregnancy. In fact, to the Chinese, this is all-important, and the pregnant woman is encouraged to spend her time in peace and tranquility, surrounded by as much beauty as possible.

## High Flyer

The Barbie doll was designed by Jack Ryan (who also happens to have been one of Zsa Zsa Gabor's nine husbands). Jack Ryan also designed the real versions of the Hawk missile and Sparrow missile. Obviously a man of varied talents.

## Lacking Nothing

Forty years after she died, Henrietta Lack's cells are still going. A cell, known to lack th chromosome which prevents tumors growing, was removed from her body. The cell reproduced itself and is still at it.

## Longest Coma

An American woman, Elaine Esposito, died just short of her forty-fourth birthday. She'd been in a coma since an operation to remove her appendix when she was six years old.

## From One Extreme to the Other

At the age of 21, Adam Rainer, an Austrian, was under 1.1 meters (3 ft 11 in) tall and was officially a dwarf. But he suddenly grew almost another 1.2 meters (4 ft), and by the time he was 51 he was a giant.

## Survival Instinct

A baby survived for 84 days inside the body of her brain-dead mother (who was kept alive on a life-support machine) before being delivered safely.

## Quick Quick Mow

A turf farmer in Ohio, USA, owns the world's widest lawnmower. It's a staggering 18 meters (60 ft) wide and it can mow an acre of grass in 60 seconds. Don't tell your dad, or he'll want one for Christmas.

## That Takes the Biscuit
The largest cookie ever made was a chocolate-chip cookie with a diameter of 24.9 meters (81 ft 8 in), containing two and a half tonnes of chocolate. Only a few calories in that!

NOW WE NEED A CUP OF COFFEE THE SIZE OF A LAKE!

## First Restaurant

Before there were such things as restaurants, food could be bought in taverns or drinking houses. The first real restaurant was opened in Paris (where else?) in 1765. It was called the Champ d'Oiseau, and it had a sign over the door that said in Latin:

"Venite ad me, omnes qui stomacho laboratis, et ego restaurabo vos"

This means, of course: "Come to me, anyone whose stomach groans, and I will restore you," which would be a good thing for restaurants nowadays to have on their signs, except they should replace the word "restore" with "ignore"!

## Illegal Substances

Tomatoes used to be grown for decoration and it wasn't until the early nineteenth century that they were actually eaten. Tomatoes and potatoes were once considered such strong stimulants that people were banned from eating them in case they got too carried away!

## Shell We Play?

An American man, Johnny Dell Foley, threw an uncooked egg 98.51 meters (323 ft) to his friend without it breaking. What we want to know is, how many did he break when he was practicing?

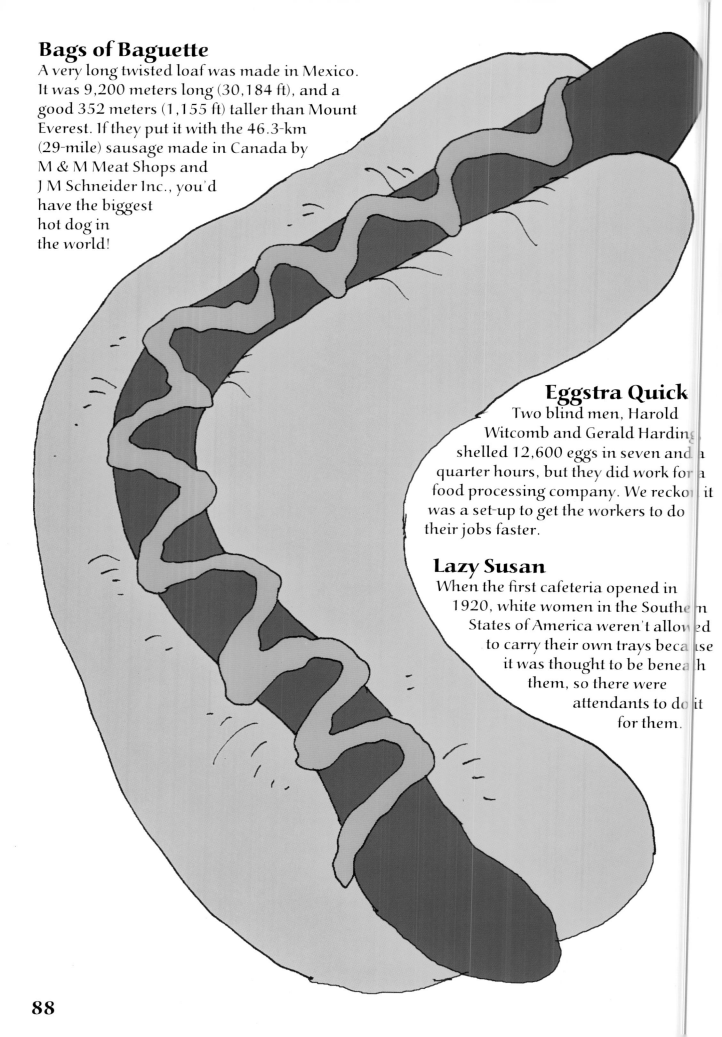

## Bags of Baguette
A very long twisted loaf was made in Mexico. It was 9,200 meters long (30,184 ft), and a good 352 meters (1,155 ft) taller than Mount Everest. If they put it with the 46.3-km (29-mile) sausage made in Canada by M & M Meat Shops and J M Schneider Inc., you'd have the biggest hot dog in the world!

## Eggstra Quick
Two blind men, Harold Witcomb and Gerald Harding, shelled 12,600 eggs in seven and a quarter hours, but they did work for a food processing company. We reckon it was a set-up to get the workers to do their jobs faster.

## Lazy Susan
When the first cafeteria opened in 1920, white women in the Southern States of America weren't allowed to carry their own trays because it was thought to be beneath them, so there were attendants to do it for them.

## In-Bread Superstitions

Did you know that it's bad luck to turn a loaf of bread on its end, and that you should never never cut a loaf at both ends, or leave a knife stuck in the bread? You shouldn't really use a knife at all, because bread should be broken not sliced. And if you toast bread on a knife, you'll be poor all your life, so the saying goes, but if you eat the crust of a loaf you'll be lucky.

## Help Yourself

The first self-service restaurant opened in New York in 1885, but only to men. However, it wasn't until 13 years later that they thought up the idea of having a tray to take the food to the table.

## Oodles of Noodles

Simon Sung of Singapore made 8,192 noodles in less than one minute.

# *WACKY DREAMS*

## Sleepless Dreams

During dreaming sleep or *REM* (Rapid Eye Movement), there is an increase in *adrenaline* (the hormone we produce in dangerous or exciting situations). The brain and heart are working at a level similar to when we are awake, so there is also an increase in the amount of oxygen we take in. This means our lungs are working faster, so we're not quite panting but we're getting there! That's why dreaming sleep is sometimes known as *paradoxical* sleep, because there's so much stuff going on, it just ain't relaxing!

90

## Bad Night

Nightmares usually happen to let us know that we've got something on our mind that we're not quite facing up to. So that great big zombie you saw last night was all about the fact that you haven't finished your homework, washed the car when you promised to, put the tortoise out or whatever.

## You Wish!

Wish fulfilment dreams are usually dreamed by children. They are dreams of getting back at that horrible kid who stole your last candy bar or of having something you've always wanted – like that pair of trainers. (Remember them? When you told your Mom how much they were, she said she only wanted to buy one pair, not the whole store!)

## Dream Position

The best sleep position for good dreaming is on your right-hand side. The other important thing for dreaming is to make sure your bed is facing in the right direction! It's true! If your bed faces north-south, sleep is more restful because of the magnetism between the two poles (North and South that is, not the curtain poles).

## Reality Shift

The men in white coats have proved that animals do have dreams of sorts. When Rover is whimpering and twitching in his basket he might be dreaming of chasing next door's cat. The only trouble is, animals can't tell the difference between their dreams and reality. So if your pet wakes up in a grumpy mood and is a bit "off," it's probably because he's had a bad dream about you telling him off or that you've taken his precious bone away.

## Sleep Tonic

Babies sleep the most, old people the least. During sleep our bodies produce a growth hormone – this only happens during our deepest slumber. The hormone repairs our bodies and in childhood helps us to grow. The more we sleep, the more we can sort out our heads. This is because plenty of sleep – and therefore plenty of dreams – helps us to work out conscious and unconscious problems (ones we might not even be aware we had). So if you want to be tall and cool, get sleeping.

# ANOTHER WACKY QUIZ

Did you enjoy that little collection? Take a minute to look at this little quiz. See how many of the questions you can answer and then try it out on your friends. I bet they won't believe that any of the answers can be right. You and I know different though, don't we!

1. **The largest cookie ever made was?**

a  Almost 25 meters in diameter?

b  Almost 250 meters in diameter?

c  Almost eaten by a hungry elephant?

YUK!

2. **The longest sausage ever made was?**

a  4.6 kilometers long?

b  Just over 46 kilometers long?

c  Disgusting?!

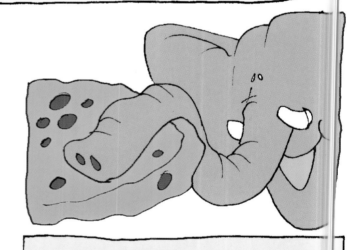

**3.** It's bad luck to do what with a loaf of bread:

a  Leave a knife stuck in it?

b  Cut it in half?

c  Stuff the whole thing in your mouth at once?

**4.  REM stands for:**

a  Random Eye Movement?

b  Rapid Eye Movement?

c  Really Eggy Monster?

**5.  If you want to be tall you should?**

a  Sleep more?

b  Stay awake more?

c  Wear heavy shoes and hang from your finger tips?

**6.  Nightmares are caused by:**

a  Disturbing noises?

b  Worries?

c  Watching too many scary movies?

Answers on page 173.

## Sea Monsters

There are already over 25,000 species of fish and each year another hundred or so species are discovered. There have been many reports of sea monsters, most of which have been pooh-poohed and the people who saw them accused of making it up. But when you consider that the oceans cover nearly three-quarters of the Earth, it becomes more believable that there are giant creatures of unknown species swimming around down there. When you realize that an ocean like the Pacific is larger than all the continents put together and that there is a trench in it called the Mariana Trench, which is 11,000 meters (36,300 ft) deep so you could plop Mount Everest down it with plenty of room to spare, then it is even less surprising to imagine that massive fish live in the oceans, mostly undetected by us snooping humans.

## Snacking Serpent

In 1962 a hissing sea serpent (try saying that quickly) attacked a life raft which had five survivors on board. It had a head a bit like a turtle with a 4-meter (13-ft) neck and apparently gave off a foul-smelling odour of rotting fish (which isn't that strange since it was one – a rotten old fish). The monster capsized the raft and took four of the men (just a snack, you understand), leaving one survivor who must have felt great for the rest of the trip.

## Monster Munch

The USS *Stein*, a frigate, was on its way from California to South America on a submarine tracking mission. Shortly after crossing the equator, the sonar gear broke down so it turned back. The ship was taken to the naval dockyard for repairs and there they found that the sonar dome was battered and full of tears and gouges, with hundreds of sharp hollow teeth stuck into it. The teeth were examined by experts, who spent months trying to find out what creature owned teeth like these. Their verdict was that it was something enormous, but as yet undiscovered by humans!

## Craven Kraken

The Vikings believed in the existence of sea monsters called "Kraken", claiming that these octopus-like creatures could overturn boats. Kraken had suckers on their tentacles and sharp beaks which could split a boat's hull apart.

## Savage Seaweed

In Newfoundland, Canada, in the late seventeenth century, two fishermen were surprised to see a large lump of seaweed heading towards their boat. The "seaweed" attacked them by wrapping tentacles around their boat. They beat it off eventually, and a chop with an ax left a 6-meter (20-ft) length of tentacle on board.

## Cryptic Creatures

The study of animals that are not recognized or accepted by other scientists is called *cryptozoology*. Some cryptozoologists think that there is a strong probability that huge 50–60-meter (165–200 ft) monster squid lurk deep in the tropical oceans, and witnesses have reported sea monsters of at least 50 meters (165 ft) in length, which backs up their theory.

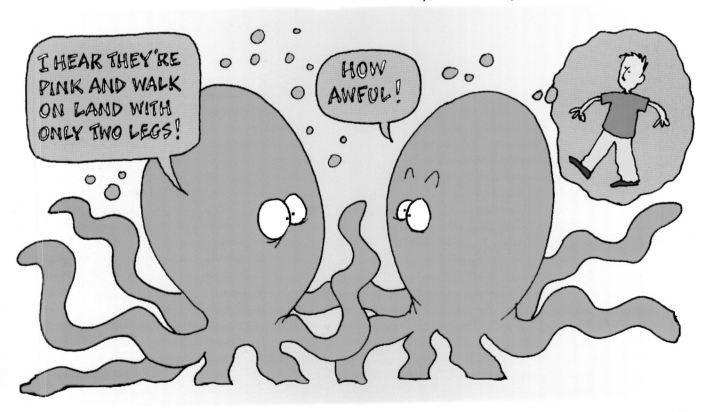

# Giant Squid

Giant squid have been described as terrifying monsters with enormous tentacles that grab their prey. The tentacles have sucker pads which hold the prey in a vice-like grip, then the squid uses its huge, parrot-like beak to tear off chunks of flesh. The beak of a squid is supposed to be powerful enough to cut a heavy wire in two. These must be the "kraken" the Vikings were talking about!

# Red Alert

Giant squid like the color red, which isn't good news if you happen to be floating around in a red life jacket or raft. There are some nasty stories of what happened to torpedoed men during the Second World War.

# Whale Fight

Squid supposedly attack massive sperm whales, and gouges that have been found on captured whales suggest that this is true. The crew of one ship watched a wrestling match between a whale and a giant squid. Both creatures reared up out of the water before sinking down again, this time with the squid in the whale's mouth. The ship harpooned the whale and it vomited up bits of squid with tentacles "as thick as a man's body."

## Close Encounter

A ship's captain had an encounter with a giant squid. He looked overboard and saw "a cold, malevolent, unblinking eye . . . coldly hypnotic and intelligent . . . which seemed to be looking directly at me," which he realized belonged to a colossal squid. The creature lay alongside the boat, which was at least 53 meters (175 ft) long, and the squid could be seen stretched from bow to stern.

## Lake Monsters

Lots of deep lakes around the world are said to have monsters and it's just possible that, as some of them were once seas and have now become isolated lakes, a few monster sea creatures have been left behind in what must seem to them like a puddle in comparison!

## Monster Bits

A giant octopus was washed up on the beach in Florida, USA, in 1896. Its body was 6.4 meters (21ft) across but the tentacles were buried in the sand and were not measured. However, chunks of the monster were chopped off and preserved. These were analyzed some time later and showed that the octopus had every chance of being over 60 meters (200 ft) long including tentacles. The largest known octopus is supposed to be one-tenth that size.

## Nessie

One of the most famous monsters must be the Loch Ness monster, or "Nessie," which is supposed to live in a loch (lake) in northern Scotland. Nessie is thought to have flippers, one or two humps and a long thin neck. He or she resembles a *plesiosaur*, a marine reptile thought to have died out with the dinosaurs about 65 million years ago. Loch Ness sits in a chain of lakes that used to be an arm of the Atlantic Ocean. It is 38 km (24 miles) long and the widest part is 2.5 km (1.5 miles). It's also 272 meters (900 ft) deep in places, so there's plenty of room for a monster (or monsters) to hide.

## Pearl Dive

One ship, the *Brunswick*, was attacked by a giant squid, which was chopped up by its propellers. The *Pearl*, a 150-tonne schooner didn't have such a lucky escape. Watched by the amazed and terrified crew of another ship, the *Strathowen*, the *Pearl* was dragged beneath the waves by the monster.

## Size Matters

Most sightings of sea and lake monsters describe the creatures as "thin at one end, fat in the middle, and thin at the other end." Joking apart, they do all seem to look rather like that. Most are covered in short sleek fur like a seal, with big eyes that reflect the light like a cat's positioned centrally on the "face." Some have been described as having a head that looks a bit like a horse or a camel, and a lot of monsters apparently have manes (which might or might not be seaweed).

## Mermaid Makers

People once believed that creatures who had the face and body of a woman and the tail of a fish really existed. Between the sixteenth and nineteenth centuries, mermaids were "made" by sewing the head and torso of a shaved monkey onto the body and tail of a large fish, and "preserved mermaids" were exhibited at freak shows.

## Pet Names

The Aborigines have got a monster called Bunyip, the Canadians have one called Manipogo, which lives in Lake Manitoba, and in New York State the Americans have one called "Champ," which lives in Lake Champlain.

## It Was This Big!

Some large fish have been mistaken for monsters. In Lake Hanas in China, enormous red fish about 9 meters (30 ft) long were seen. They had "heads the size of car tires" and were swimming in large shoals.

## Mini-Monsters

Other lakes have their own mini-monsters. Catfish are known to grow to great lengths (to get themselves noticed), and can often measure up to 5 meters (16.5 ft). A Russian sturgeon caught in Europe's longest river, the Volga, was 22 meters (24 ft) long and weighed 1,473 kilos (3,242 lb)!

## Radioactive

There's another theory that these monsters were actually normal-looking fish until they started eating all the nuclear waste that we humans chuck into the sea. Now they've mutated, and the results are some dreadful monsters that are hatching up a plot at this very moment to take over the world. (This is all nonsense of course.)

# WACKY MONSTERS QUIZ

Did you enjoy that little collection? Take a minute to look at this little quiz. See how many of the questions you can answer and then try it out on your friends. I bet they won't believe that any of the answers can be right. You and I know different though, don't we!

1. **Three-quarters of the Earth's surface is covered by:**
   a  Water?
   b  Land?
   c  Carpet?

2. **Cryptozoology is the study of:**
   a  Extinct animals?
   b  Animals not recognized or accepted?
   c  Spooky crypts?

**3. Giant squid are known to like:**

a  The color red?

b  The color yellow?

c  Going to Spain for their holidays?

**4. Loch Ness is:**

a  In Wales?

b  In Scotland?

c  A home for retired sea monsters?

**5. A mermaid has:**

a  The body of a woman and the tail of a fish?

b  The body of a fish and the legs of a woman?

c  A terrible time with her makeup?

**6. A sturgeon caught in the river Volga was:**

a  22 meters long?

b  2.2 meters long?

c  Very annoyed?

Answers on page 173.

## Ghost Ship

*Lady Lovibond* ran aground off Goodwin Sands, off south-east England, on 13 February 1748 with a wedding celebration on board. In 1798, 50 years later to the day, a three-masted schooner identical to the *Lady Lovibond* was seen heading towards the Goodwins. Witnesses heard the sounds of laughing voices and then terrified screams as the ship ran aground. On both 13 February 1848 and 1898 the same ghostly sight was seen.

## Flying Dutchman

The *Flying Dutchman* was captained by Hendrik van der Decken who was known to be a little odd. On his last journey from Amsterdam to the East Indies, he apparently made a pact with the devil as his ship rounded the Cape of Good Hope. He was condemned to sail on until the Day of Judgment. The *Flying Dutchman* – or a ship closely resembling it – is still seen roaming the seas. Sighting the ship is supposed to bring a curse or death on the person who sees it.

## Bad Ship Day

There is a superstition among seafaring folk that a Friday is not a good day to launch a ship. One of the reasons Friday is a bad day generally is that Jesus was crucified on a Friday, but on that basis, if you were a Christian, you wouldn't do anything on that day, which isn't practical.

Anyway the British Navy decided to ignore this, in order to prove a point, and launched a ship called *HMS Friday* on a Friday. And they even managed to find a captain called Captain Friday. Unfortunately it sank with all lives lost, so all they proved was that you shouldn't launch a ship on a Friday.

## The *Ourang Medan*

A Dutch freighter, the *Ourang Medan*, was sailing to Indonesia in February 1948, when an SOS call was heard. When rescue ships reached it, all on board were dead, with their eyes wide open in horror and their hands raised up as if to push something away. Even the ship's dog had died like this. There was no sign of poisoning or any kind of disturbance. What happened to the people aboard the freighter remains a mystery to this day.

# Mystery of the *Marie Celeste*

On 3 December 1872, an American ship, the *Marie Celeste,* was found drifting off the Azores. The ship's captain, his wife and two-year-old child, and a crew of seven were missing. The captain's log was normal, the last entry being 25 November, and there was no sign of mutiny, although one of the lifeboats was missing. Everything else was shipshape and breakfast had even been served. The question is, how did the *Marie Celeste* travel 800 km (500 miles) on her intended course, for ten days, with no one at the wheel to steer her?

## The Philadelphia Experiment

An experiment in invisibility was made by the US navy in 1943. The Philadelphia Experiment was, and still is, top secret. The idea was to use electromagnetism to try to make ships disappear, in order to protect them from enemy attack in the Second World War.

In the naval dockyard in Philadelphia, the USS *Eldridge* (and its crew) were subjected to immensely powerful, pulsating electromagnetic fields. The results were amazing. First of all a greenish mist appeared. Then the ship gradually disappeared, but was seen by witnesses to appear in the dockyard in Norfolk, Virginia, over 320 km (200 miles) away! It stayed there for a couple of minutes, and then rematerialized in Philadelphia.

*Terrible Side Effects*
Although there have been no reports made available to the public, it is known that many of the crew became ill after the experiment. Some went mad and others kept disappearing, two of them never to return again; two more men spontaneously combusted (suddenly caught fire) and burned to death. Sounds like a tall ship story? We shall see. Or will we?!

# Dawn Disappears

In the 1930s a 75-tonne schooner called *Dawn* went missing on its voyage from Alabama, USA, to Barbados. It had an experienced captain and a crew of eight on board, and the weather conditions were good. An air and sea search was carried out but nothing was found. Three months later the ship was found off the coast of Mexico. It was in a good condition, with half a tank of fuel and the sails neatly furled. But the captain and crew had completely disappeared.

# The Day the *Queen* Didn't Turn Up

What happened to the *Island Queen* is similar to the mystery of the *Marie Celeste*, but even more strange. In August 1944, the schooner left Grenada for her 120-km (75-mile) overnight journey to St Vincent. She had on board 57 tourist passengers and an experienced captain and crew.

Three hours after the *Island Queen* set sail, another smaller ship, travelling the same route, caught up with the schooner, and the crew of that ship heard laughter and singing on board the larger vessel. At midnight, the smaller ship (which could sail closer to the shore) moved away from the *Island Queen* and continued its course to arrive in St Vincent the next morning. The *Island Queen* never arrived. Nothing of the ship or its passengers was ever seen again.

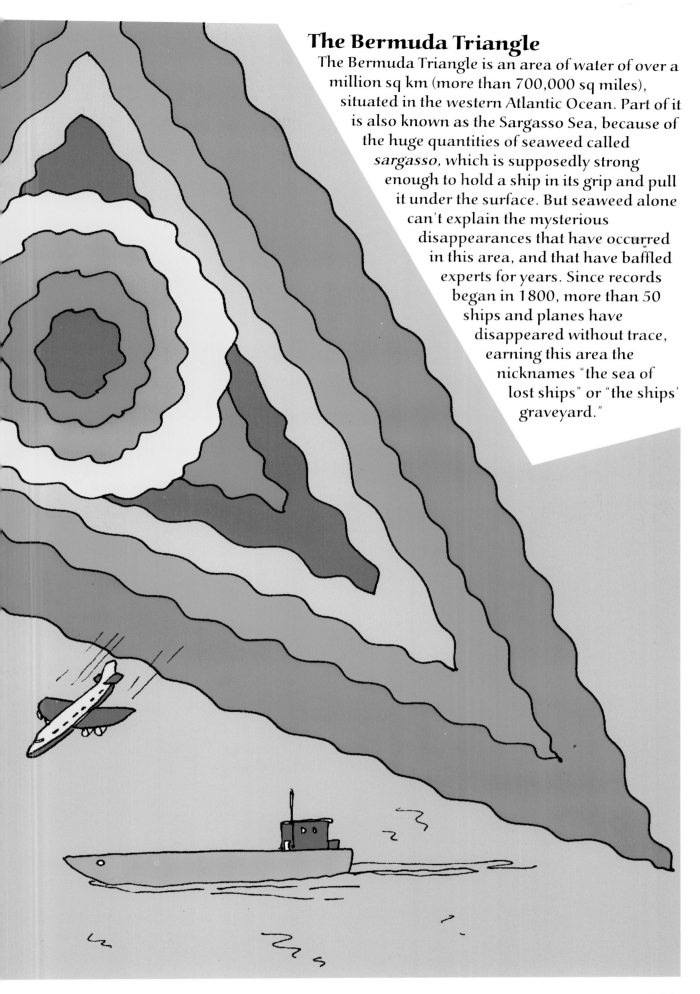

# The Bermuda Triangle

The Bermuda Triangle is an area of water of over a million sq km (more than 700,000 sq miles), situated in the western Atlantic Ocean. Part of it is also known as the Sargasso Sea, because of the huge quantities of seaweed called *sargasso*, which is supposedly strong enough to hold a ship in its grip and pull it under the surface. But seaweed alone can't explain the mysterious disappearances that have occurred in this area, and that have baffled experts for years. Since records began in 1800, more than 50 ships and planes have disappeared without trace, earning this area the nicknames "the sea of lost ships" or "the ships' graveyard."

117

## Abandoned Ships

In some cases, it is only the crew who disappear, abandoning the ship for no apparent reason; the ship is found adrift and usually surprisingly intact, some time later.

In 1881 the *Ellen Austen* was passing the Azores when the crew saw an abandoned schooner. Three men went aboard, but a storm separated the two boats and when the *Ellen Austen* came upon the schooner again a few days later, the men had disappeared. Three more men went aboard, but again a storm separated the two ships. The schooner, and those of *Ellen Austen*'s crew who had gone aboard, disappeared for ever.

In 1902 a German ship, the *Freya*, was found adrift with no crew, and in 1932, the *John and Mary* was also found abandoned in perfect weather conditions. In both cases, the crew were never found.

The USS *Cyclops* went missing during the First World War. In March 1918, the supply vessel was travelling from Barbados to Virginia, USA, when all radio contact was lost. There was no enemy presence found to have been in the area, and an air and sea search failed to find anything remaining of the USS *Cyclops*.

In 1963, a massive freighter called the *Sulphur Queen* disappeared without trace, as did the *Anita* and her 32-man crew, ten years later.

## Missing Planes

Many planes have gone missing in the area too, but it wasn't until 1945, when five planes vanished while on a routine training flight, that the Bermuda Triangle was first named as such, and when the world suddenly became alerted to the weird happenings in the region. Because the flight path of the doomed planes formed a triangular route, which pointed to the island of Bermuda, the area became known as the "Bermuda Triangle," and the name has stuck.

ARE YOU SURE THAT'S BERMUDA DOWN THERE

Two identical passenger aeroplanes from the same airline, Star Tiger and Star Ariel, disappeared in 1948 and 1949 respectively, one from within the Bermuda Triangle and the other a long way outside it. Both pilots gave a radio message reporting good weather conditions and no problems, but in each case it was to be their last. A massive air and sea search was carried out, in which a total of more than a million miles was flown by US military planes, but no wreckage or bodies were ever found. Did both aeroplanes have the same fault or weakness which caused them to crash into the sea? Or did they fly into another dimension . . . ?!

## Still Swallowing

Even with the aid of modern technology, no one can explain the strange power of the Bermuda Triangle to swallow up ships and aircraft.

*THE COMPASS HAS GONE CRAZY!*

In 1966, Captain Don Henry was taking a tug to Fort Lauderdale, USA. In calm seas and good weather conditions, the tug's compass began to spin madly and a milky cloud loomed towards the tug. Suddenly all electrical power on the boat was lost and it felt as if the tug was being pulled towards the cloud. Captain Henry had to use full throttle to pull away from it, but what seemed like a magnetic force was so strong that the tug only just managed to get clear.

## Unsolved Mystery

What is behind the disappearances and strange happenings in the Bermuda Triangle? Is there a huge magnetic force that pulls the ships downwards, or perhaps violent underwater currents that suck ships to their doom? Could some of the missing ships and planes have been caught in a waterspout, caused by a funnel of air that rushes down to the ocean from storm clouds above?

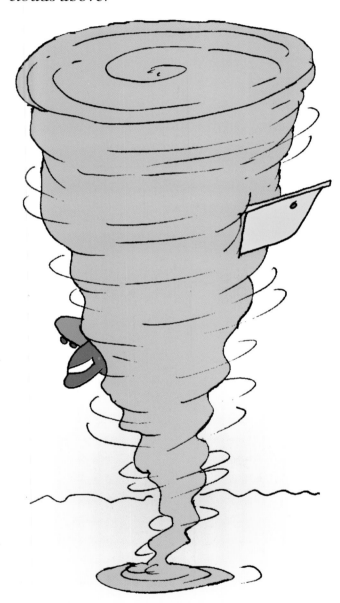

When the funnel reaches the sea, it sucks up water and anything in its path is tossed into the air, then dumped on by the massive volume of water that is eventually released from the spout. Or perhaps the ships were wrecked by a freak storm, the bodies of crews eaten by sharks and sea monsters, and the ships' carcasses sucked into holes in the sea bed or in the sides of islands, never to be retrieved.

Could it be that UFOs have destroyed the ships and planes, and kidnapped their crews? Or do the remains of an unknown, ancient but advanced civilization lie beneath the ocean? The first question we must ask is: Will *we* ever know? The second question is: Do we *want* to . . . ?

# WACKY SHIP STORIES QUIZ

Did you enjoy that little collection? Take a minute to look at this little quiz. See how many of the questions you can answer and then try it out on your friends. I bet they won't believe that any of the answers can be right. You and I know different though, don't we!

1. **The Flying Dutchman is:**

a A ghost ship doomed to sail the seas forever?

b A ghost ship abandoned by her crew?

c A striker for Manchester United?

2. **The captain of HMS Friday was named:**

a Saturday?

b Friday?

c Dennis?

**3.** The Philadelphia Experiment was performed on the:

a  USS Philadelphia?

b  USS Eldridge?

c  USS Marie Celeste?

USS MARIE CELESTE

HAVE YOU SEEN DONALD?

**4.** The Bermuda Triangle is located in:

a  The Pacific Ocean?

b  The Atlantic Ocean?

c  A duck pond?

**5.** The Bermuda Triangle was first named in:

a  1945?

b  1954?

c  The Jurassic era?

I NAME THIS TRIANGLE...BERMUDA

**6.** A possible explanation for the Bermuda Triangle disappearances is?

a  Waterspouts?

b  Volcanoes?

c  A giant plug hole?

Answers on page 173.

# WACKY SUPERNATURAL WORLD

## Day Glow

There are many weird things that happen to people which cannot be explained by science, like the strange human condition of *chromidrosis*, where sufferers perspire colored sweat. Green, purple, yellow and black are among the colors known to have been produced by people with this condition. Imagine, you could color co-ordinate your sweat with the clothes you wear!

## X-Ray Eyes

Not just something out of science fiction, there are actually people who have X-ray eyes. A Ukrainian woman can see other people's insides through their skin, and she now works in a hospital diagnosing diseases.

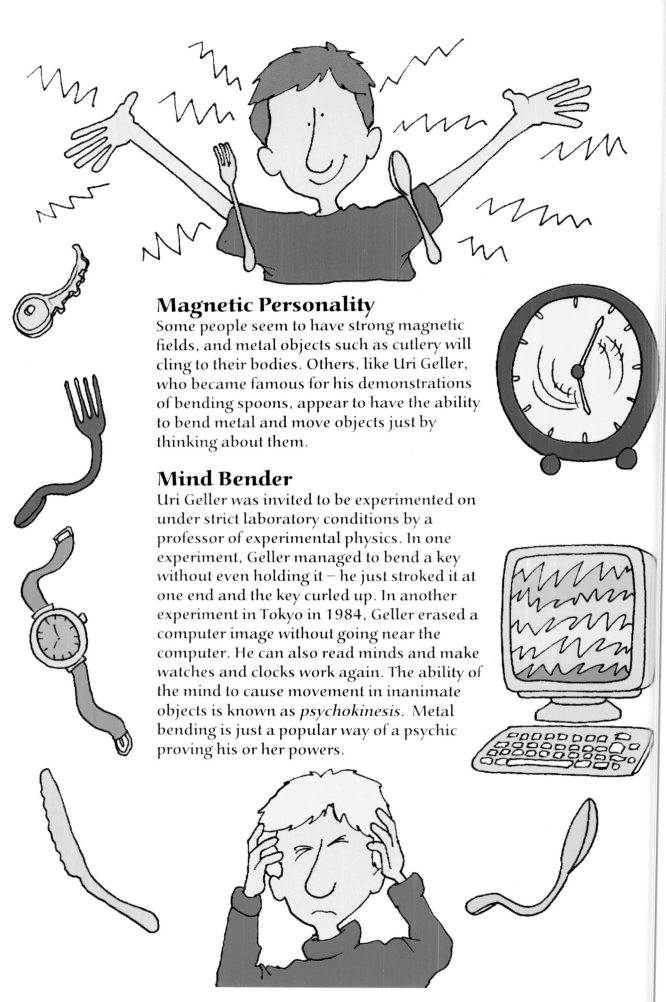

## Magnetic Personality

Some people seem to have strong magnetic fields, and metal objects such as cutlery will cling to their bodies. Others, like Uri Geller, who became famous for his demonstrations of bending spoons, appear to have the ability to bend metal and move objects just by thinking about them.

## Mind Bender

Uri Geller was invited to be experimented on under strict laboratory conditions by a professor of experimental physics. In one experiment, Geller managed to bend a key without even holding it – he just stroked it at one end and the key curled up. In another experiment in Tokyo in 1984, Geller erased a computer image without going near the computer. He can also read minds and make watches and clocks work again. The ability of the mind to cause movement in inanimate objects is known as *psychokinesis*. Metal bending is just a popular way of a psychic proving his or her powers.

# Reincarnation

Some of us believe that we've been here before and perhaps have a strong connection with certain periods of history, like Dorothy Eady, who, after a fall in 1907, began to have dreams about life in ancient Egypt where she was a queen. Some people think they might even have been animals in a past life, or will become one in the next. It makes you think twice before swatting that irritating fly!

*Here are some wacky reincarnations:*

| Is | Was |
|---|---|
| Canoeist | Otter |
| Potholer | Mole or worm |
| Aviator | Bird |
| Swimming Champ | Fish |
| Sprinter | Cheetah |
| Porter | Mule |
| Logger | Beaver or Elephant |
| Builder | Termite |
| Florist | Bee |

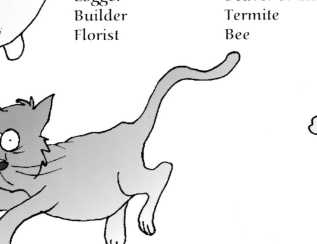

## Blast from the Past

*Psychometrists* are people who can touch an object belonging to someone else and pick up emotional vibes, allowing them to feel what the owner was like or where the item was made. In 1917, a battered suitcase was found on a beach in Ireland by an amateur psychometrist. When he picked it up he felt as if he was suffocating. He opened the suitcase and traced the owner – it had belonged to a passenger who had drowned in the sinking of the liner *Lusitania*, two years earlier.

## Life After Death

Two Italian scholars, Marsilio Ficino and Michele Mercati, often discussed whether or not there was life after death. One day they agreed that whoever died first would try to come back to visit the other. Early one morning in 1491, Mercati heard a horse galloping down the road. It stopped outside his door, and he heard his friend Ficino shouting to him that what they had said about the other world was true. Before Mercati could stop him, Ficino galloped away on his white horse. He found out later that his friend had died at around the time Mercati had seen him on the horse.

## Banana Splits

Some people seem to have a split personality or even multiple personalities or are they just bananas? A woman called Christine Sizemore had 23 different personalities over the space of 20 years. She could be a really nasty character or a quiet holy one, and anything in between (does this sound like anyone you know?!). Another woman called Sybil Dorsett had 16 different personalities, including a quiet girl, a religious freak, and two different men. She often thought she was a county in England too (only joking!).

## Virgin on the Ridiculous

Hundreds of sightings of statues and other religious images weeping blood and tears have been recorded throughout the world. They are usually statues or paintings of the Virgin Mary or Christ. Often the tears are found to be the same stuff as human tears, which disproves the theory that the tears are just water leaking out of the plaster. In New York in 1960, a picture of the Virgin Mary wept real tears, but these vanished when they reached the bottom of the picture frame. In Italy in 1971, a man woke to find a painting of the Madonna dripping blood from under the glass of the painting. Blood was coming out of her eyes, heart, hands and feet. Police took the painting away and placed it in a locked box, but when they looked the next morning the painting was bleeding again – and it was found to be human blood!

**129**

## Dead But Not Rotten

There have been cases where the bodies of people, usually saints, have not decomposed (rotted) as they should. St Charbal Maklhouf, whose remains are kept at the Monastery of St Maro, in Lebanon, is one such saint. After his burial a bright light surrounded his tomb for 45 days! When he was exhumed (dug up) in 1899, his body was perfectly preserved, but seemed to be seeping a watery blood. The body was buried again for 23 years, and in 1950, pilgrims noticed liquid leaking from his tomb. It was exhumed again, and found still not to have decomposed. Now how do you explain that?

## Strange Faces

Some of the leaking blood forms itself into images on the wall beside the statue or picture; these images are often in the form of the face of Christ or the Virgin Mary, and sometimes other saints. Most experts cannot understand how these extraordinary events take place, but agree that in many cases, no sign of trickery is present, therefore the bleeding and weeping must be caused by poltergeist phenomena.

# Wacky Miracles

## Lourdes

In 1858, a young girl, Bernadette Soubirons, had several visions of the Virgin Mary telling her to uncover a spring at a place called Lourdes in France. She did so, and it has been a place of healing ever since, with thousands of people claiming to have been miraculously cured.

HE'S CONVINCED IT WILL CURE HIS HEADACHES!

## Knock Knock

The Archbishop of Hobart, all the way down there in Tasmania, rubbed cement in his eyes and, would you believe it, his sight returned. The cement came from the church at Knock in Ireland, which is another holy healing site.

## Shamrock Shocker

In 1929, a sixteen-year-old girl called Mary Collins, from Southern Ireland, told her sister that she was going to die that night. In the morning she was dead and was duly buried. However, when her grave was dug up in 1981 to make room for other relatives, her corpse had not rotted and her eyes were wide open and "shining like diamonds." Since then pilgrims have come from around the world and many people claim to have been cured by visiting her grave.

MARY COLLINS

I WAS ONLY KIDDING!

## Nuts

In the UK in 1979, one Southampton couple were showered with beans every time they opened their door, as were people who came to visit them. Another couple, Mr and Mrs Osborne of Bristol, UK, were showered with hazelnuts out of a completely clear sky. There were no buildings above them from which the nuts could have been thrown, and no mischievous squirrel anywhere to be seen. Where the hazelnuts came from is a mystery, but apparently they were very tasty.

# Raining Cats and Frogs

There are lots of reports of strange things just dropping out of the sky, usually when the weather is clear and the sky blue. It is understandable to find that a lump of frozen airplane toilet waste has just gone through your roof, but most items are too mundane or ridiculous to have been made up. Beans, mustard and cress, frogs, newts, ducks, lizards and fish have all been reported. In fact, a shower of frogs fell on a village in the Midlands, UK, in 1944. The village name was Hopwas!

# Mind Over Matter

If you really believe you can lie on a bed of nails without being harmed (punctured!), then you can, or so they say, but don't try this at home. Some people believe that they can handle poisonous snakes without being bitten. One devout evangelist decided to take the Bible literally – especially the bit that says, "Thou shall pick up serpents." All those who followed him were encouraged to test their faith by handling venomous snakes. The snakes were first put into a box which was kicked around a bit in order to make the snakes really angry, then they put their hands in the box. The holier the person, the less likely they were to be bitten. Mmm. The cult became so popular that insurance companies in the USA no longer treat death from a snake bite in church as accidental!

# Faith Healers

In 1900 a 23-year-old travelling salesman called Edgar Cayce suddenly became mute. It's a bad idea not being able to speak if you're a salesman, so he naturally sought medical assistance. When that was no help he turned to unorthodox medicine. He consulted a hypnotist who discovered that Cayce had an extraordinary resistance to suggestion, which meant that the hypnotist couldn't put ideas into Cayce's head – Cayce would have to do it for himself. The hypnotist put Cayce into a trance and told him to cure himself because nobody else would be able to. It worked!

The hypnotist was convinced that he and Cayce could work together to heal others but Cayce refused. Immediately he lost his voice again, which convinced him to take up the hypnotist's offer. The hypnotist would put him into a trance, and he would heal people who came to see them. Soon though, he didn't even need to see the patients, he just needed a list of names and addresses and he would be able to suggest treatment.

## It's the Real Thing!

Sometimes Cayce would tell a patient to eat a certain kind of fruit in great quantities and drink a lot of Coca-Cola! When he was in a trance he changed from being a not particularly clever person to a man of great knowledge. By 1911, he was well known as a spiritual healer. He could also tell the future and find missing items such as murder weapons!

## Either he Goes, or I Do . . . !

Cayce told the president of a railway company that if the president didn't dismiss one of his staff, that man would be responsible for a major accident on the railway. The president ignored him, the accident happened, and the company president was killed.

## Wacky Disappearances

In 1926, South American Carlo Mirabelli disappeared from a railway platform in Sao Paulo, Brazil. He had been standing with a group of friends who all witnessed the disappearance, as did other people waiting in the station. To his shock, he turned up two minutes later, some 90 km (56 miles) away, close to where he'd been planning to go with his friends that day. Mirabelli's ability to *teleport* himself became well known. On one occasion, he underwent a test where he was tied to a chair in a locked room with five witnesses watching him. He turned up in a room next door a few minutes later.

## A Nose for It!

Kings and queens were supposed to have healing powers and one touch from royalty would cure many ailments. In the early seventeenth century, King Charles II of England was walking in St James's Park in London, UK, when he was accosted by a mad Welshman called (of all things) Arise Evans. The man had a severe "fungus" growth on his nose, and Evans had dreamed that if the king's hand were to touch his nose, the growth would disappear. As you can imagine, King Charles wasn't too sure about this, but the Welshman grabbed the king's hand and rubbed his nose with it. And, miraculously, he was cured (although it's rumored that the king's hand turned green for a while . . .).

## Min-Min

An oval-shaped fluorescent light, known as the Min-Min Light, is often seen rolling across the sky near Warenda Station in Western Queensland, Australia. No one has been able to explain where the light comes from or where it goes, although in 1917 it was seen coming from a cemetery by a man who followed it on his horse.

# Wacky Appearances

You know what it's like when something suddenly turns up that you lost a long time ago. Usually it's where you dropped it, or hidden behind a piece of furniture or clothing. But in some cases, items have been lost a long way away from where they are eventually found, with no explanation for their sudden appearance.

## *Flying Knife?*

A Bristol merchant went to sea, leaving his wife and children at home. One day on board ship he dropped his knife over the side by mistake, and he resigned himself to having lost it for ever. But at that same moment, at his home in Bristol, his knife flew through the window and stuck in the floor, close to where his wife was sitting. Unbelievable eh?

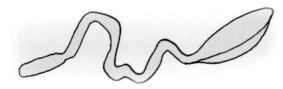

## *Ring of Fire*

The author Wellesley Tudor Pole was shot at while sailing in Egypt in 1918. He wasn't hit, but in the ensuing kerfuffle lost a treasured ring in the waters of the Nile. Three months later, when he was back in Cairo at his desk, he was shot at again from a building across the street. The bullet missed him and embedded itself in the opposite wall. At the same moment, his lost ring dropped down in front of him. The circumstances of the gunshot were present at both the losing and the finding of the ring, but the rest of it is inexplicable.

## Hologram Homes?

A set of cottages near Chagford in Devon only appear on the moor every ten or twelve years, but some people have been inside them and have even talked to their occupants. Where do you think they go when they disappear?

## Unreal Estate

One day in 1931, a young girl cycled to see a friend who lived some miles away. A storm blew up and it began to rain heavily. The girl saw a pretty cottage with smoke coming from the chimney and surrounded by a well-tended garden. She stopped at the cottage and asked for shelter from a kind and smiling old man who welcomed her in. Later she found herself outside and cycling once again, although the rain had stopped and she was warm and dry. When she told her friend that she had sheltered at the cottage, her friend told her not to be silly, that there was only a derelict cottage on the road she had taken. On her way back, the girl saw that her friend was right – the cottage where she had sheltered was a crumbling ruin with the roof falling in and a garden full of weeds.

138

## Thin air

French-Canadian Jean Durrant could disappear in front of friends and reappear somewhere else. His friends were so astounded that they forced him to put himself under test conditions. He agreed to be tied up and locked in a room with no windows. Several times he would turn up behind his friends waiting by the door. Unfortunately, he performed the stunt once too often, and disappeared for good.

## Burning Ambition

Some people, such as Nathan Coker of Maryland, USA, appear to be fire resistant. In 1871, he was asked by a committee of rather boring people to demonstrate his abilities. They heated a shovel until it was red-hot and then placed it on the soles of his feet. He wasn't hurt, or burned, and he went on to pick up glowing coals and to pour molten lead onto his hands in front of all the witnesses.

## Electrical Faults

Others cause electrical equipment to go wrong just by being near it. One girl was seen to have a blue light flashing in her chest and a boy's hands were seen to glow whenever electrical appliances went wrong, which they always did when he was around.

## Automatic Art

"Automatic" artists can draw, paint, write or compose music in the style of long-dead artists. All you need to do is to go into a trance to have a Picasso at your fingertips. But how does it happen? Is it the subconscious mind at work, or has the person really been taken over by the spirit of the artist? In the case of written words, often the language is completely unknown to the "automatic" writer, and is sometimes in ancient script which is no longer used today.

GIVE ME ALL YOUR SWEETS OR I'LL COME ROUND AND PLAY WITH YOUR COMPUTERS!

## Spooky Sites

There are at least 16 ghosts at Bramshott in Hampshire, UK. Sometimes the ghost of Boris Karloff, the actor who played Frankenstein's monster, appears here, but even if you don't see the phantom actor, you might smell pipe tobacco burning or hear a ghostly piper play. Among other ghostly specters are a long-gone pig and an enormous ghoulish cat.

There's a big manor house on the Isle of Wight called Knighton Gorges, which was knocked down a long time ago. Sometimes at dusk on New Year's Eve the house reappears, often with all the lights shining in the windows and the sound of music playing.

# Hot Stuff!

Spontaneous Human Combustion, or SHC as it is known, occurs when a person suddenly bursts into flames. It sounds incredible, but it does happen. Normally only a piece of limb, usually a leg or foot, is all that is left, but no bones apart from that. What is left behind is an oily yellow grease, which is melted body fat and oily soot, but there is usually no smell of burning.

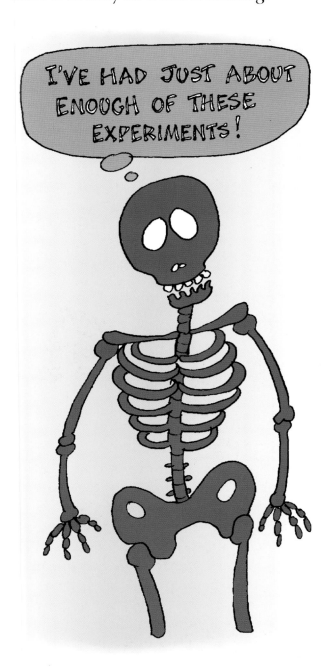

About four people are known to have spontaneously caught fire while in their cars. Although most have been burned so badly that they needed hospital treatment, and some have even died, the weird thing about it is that the car's interiors were hardly damaged. For example, one woman who caught fire in her car was badly burned, but her white leather seat was only a little scorched. The extraordinary thing is that in experiments that have been done on burning corpses in a crematorium, even those that have burned at 1,093 degrees centigrade (2,000 °F) for over eight hours still leave bones. And if the body did reach that temperature, why didn't the whole house catch fire? Most people who have burned to death in this strange way have had a ring of about 1.2 meters (4 ft) around and above them which has been burned, but outside that there is no burning or scorching.

But don't worry too much about bursting into flames: it's very rare and, funnily enough, usually happens after people have had quite a bit of alcohol to drink and a cigarette.

## Fiery Looks

Some people can make things catch fire just by looking at them, and one house had 200 fires break out in just one week. Scorch marks appeared on the walls and ceilings, upholstery and bedding caught fire. Sometimes the person who seems to "cause" the fire gets burned too, at other times they are able to touch the flames and remain unburned and unhurt.

## Spooklights

There have been "spooklights" at a place called Marfa in Texas, USA, for most of this century. They're big ball-shaped blobs of light that jump around all over the place, and if you go near them they jump away as if they're teasing. No one knows what they are or where they come from. Creepy, huh?

## Wacky Phenomena

Eclipses of the sun and moon, shooting stars, etc, were all things that were once thought of by ancient people as being bad omens. Now we know what causes them, they don't scare us any more – well, not much anyway. On the other hand, there are some things that happen which are explained away by the "authorities" as being caused naturally, which might just have something behind them.

## Laser Blazer

There are some weird things that occur naturally. These are called *phenomena* and one such *phenomenon* is the Northern Lights (known as *aurora borealis*) in the Northern Hemisphere, which are like a enormous sky-wide laser display, often in color. They are caused by tiny bits (*particles*) which have fallen out of the sun, hitting the Earth's atmospheric gases.

## Meteorite or What?

In 1908, something huge took out 1,994 sq km (770 sq miles) of Siberian forest. Shock waves from the explosions were felt thousands of miles away. Whatever it was didn't leave a hole in the ground (crater) but all the trees were flattened and burned by an enormous blast. Could it have been a meteorite or a comet? Or was it a UFO?

## Swimming Ghosts

In December 1929, two men died after they had inhaled poisonous fumes in the engine room of the tanker, *Waterton*. The day after they were buried at sea, their crewmates saw them bobbing in the waves of the Pacific Ocean, laughing and waving. On the *Waterton*'s next voyage, they were seen again, and a crew member took several photographs that show them clearly, and that the men's relatives identified as their kin.

## Skimming Saucers

A pilot called Kenneth Arnold was the man to coin the term "flying saucers" in 1947. He came across nine of them when he was flying his plane over the Cascade Mountains in Washington, USA. Being a pilot, Arnold was able to calculate their speed and he estimated it at 2,735 km (1,700 miles) per hour, so they weren't hanging about. He told reporters that the UFOs looked like saucers skimming across water and the expression stuck.

## Oh Yeah?

There were a lot of sightings of UFOs in the 1950s and 1960s and people got a bit over-excited about them. In 1952, one man took a picture of a circular vacuum cleaner top and pretended that it was a spacecraft. The funny thing is, a lot of people believed him! Another man took a close-up photograph of a button, which looked remarkably like a flying saucer, and again a lot of people were convinced that it was a spacecraft. Many people pretended (and still do) that they'd seen a spacecraft or they'd had an encounter with a ufonaut, just so they could get their name in the newspapers or on television. This is known as a *hoax*.

There are plenty of strange things going on that we think are UFOs or ufonauts but can often be explained either by natural phenomena or by hoaxes.

WHERE ARE MY BUTTONS?

TAKE ME TO YOUR LEADER

**145**

## Eerie Eavesdropping

When the space shuttle *Discovery* went up in 1988 a radio ham is supposed to have overheard a report from the astronauts that they'd seen a spacecraft.

In 1951, another astronaut, Gordon Cooper, saw "double lenticular-shaped" craft (typical shape of a flying saucer). Cooper and other pilots saw several hundred of these craft which were flying in formation, "higher and faster than any other plane of the day."

It is rumored that the film, *Close Encounters of the Third Kind* might have had government backing, because it was used as a means to condition the public for an eventual acceptance of extraterrestrials.

## UFOs

Frederick Valentich, an experienced pilot, disappeared in his Cessna light aircraft over the Bass Strait in 1978. He radioed air traffic control to say that he could see a large aircraft with four bright landing lights, which he said looked metallic and shiny. He said that it seemed to be playing some sort of game with him, then complained that the Cessna's engine was beginning to play up. His last words were, "It's hovering and it's not an aircraft." Metallic scraping sounds were heard over the radio and then nothing. Neither Valentich's body nor his aircraft were ever found.

## Ufonauts

UFO stands for Unidentified Flying Object, in case anybody wasn't sure! The people who fly them are called *ufonauts* or *extraterrestrials* (after all "alien" is a bit unfriendly).

There are ancient cave drawings of what look like astronauts and spacecraft, which suggests that extraterrestrials might have been visiting our planet for a pretty long time. There are some people who think that creatures from outer space helped build the pyramids and put the stones in a circle at Stonehenge.

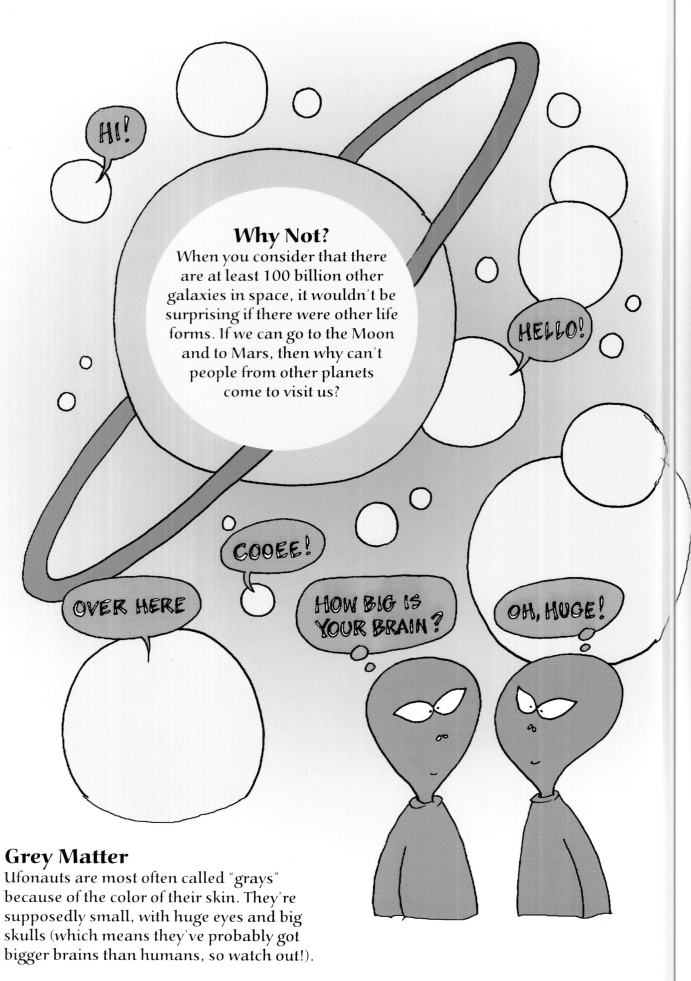

## Why Not?
When you consider that there are at least 100 billion other galaxies in space, it wouldn't be surprising if there were other life forms. If we can go to the Moon and to Mars, then why can't people from other planets come to visit us?

## Grey Matter
Ufonauts are most often called "grays" because of the color of their skin. They're supposedly small, with huge eyes and big skulls (which means they've probably got bigger brains than humans, so watch out!).

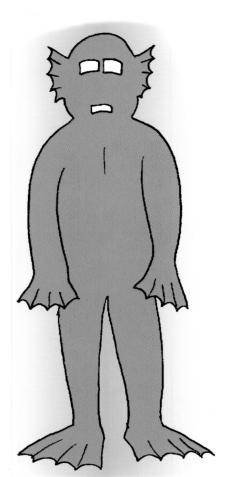

## All Shapes and Sizes

However, extraterrestrials have been described as being anything from 0.3 meters (1 ft) to 3.3 meters (10 ft) tall. Some look very similar to humans, some nothing like. As well as gray skin, some have greenish-gray and sometimes very bright green skin. Others glow, have big eyes, a slit for a mouth, no nose to speak of and webbed feet with or without talons. Some seem to be friendly or playful, others aggressive.

Some look, or can make themselves look, very similar to humans. A taxi driver in Mexico spent the night talking to two slightly odd people who appeared when his car broke down. In the course of their conversation, they told him that they were "not of this planet" and at dawn he watched them get into their spacecraft.

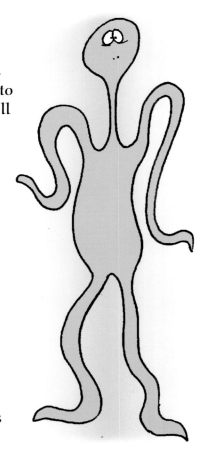

149

## Men In Black

Who are the men in black? Some people who've been abducted, or who have talked or written about UFOs, claim to have been "visited" by men in black suits whose eyes light up like lamps.

They usually appear in a group of three and are often male. If they arrive by car it is usually an expensive one but the model is often a little out of date although the car smells brand new. In some cases though, the car has a licence plate that has yet to be issued.

They're polite and not aggressive but, nevertheless, somewhat frightening, and they speak in an old-fashioned way as if they've learned the language by watching an old film. They dress all in black but with white shirts, and everything about them is slightly odd. Their clothes look brand new and they seem unfamiliar with everyday objects.

WE'LL TAKE TWO COKES... IT'S THE REAL THING

## Extraterrestrial Activities

Extraterrestrials have often been seen collecting soil or water samples and have approached humans to ask for water, presumably to drink. They have also been seen doing strange things, like measuring animals. Four poachers in Cheshire, UK, watched in amazement as a UFO landed in a meadow and four silvery men tried to measure a cow!

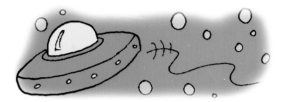

## Eggsactly What?

In New Mexico, USA, in April 1964, a police officer saw a flash of light in the sky. When he went to investigate, he saw beneath it a huge shiny egg with four metal stabilizers that looked like legs. Beside the spacecraft stood two little ufonauts. The spacecraft roared off when he got closer but it had left marks in the ground from the metal legs. A year later in France, a farmer saw exactly the same thing in his fields.

## Are They Friendly?

In many alleged encounters with extraterrestrial beings or ufonauts, the *experiencer,* as the human who has made contact with a ufonaut is called, has sometimes been paralyzed by a beam of light, but afterwards suffers in no other way. Other experiencers have felt sickness and tingling for several days after an encounter.

However, there have been a few cases where the ufonauts were aggressive. In 1967 a Brazilian farmer saw a UFO in his fields. There were three ufonauts in tight yellow suits, jumping about like children. When they saw him they ran towards him but the farmer shot at them with his rifle. Suddenly a beam of green light came out of the spaceship. He suffered sickness and tingling and was diagnosed with leukemia (a cancer of the blood). Within two months he was dead from radiation poisoning.

Another man waved to some ufonauts he saw in the Everglades in Florida, USA. A beam of light smacked him in the forehead. He was out cold for several hours and afterwards was partly blinded by internal bleeding in his brain.

## Zapped!

One man touched a UFO that had landed near a lake in Canada. His shirt burst into flames and a grid pattern of burn marks formed on his chest as if he'd been hit by a strange heat-throwing gun.

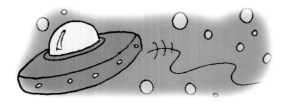

## Take That, Four-Eyes!

Another man in Missouri, USA, saw a UFO overhead as he was driving his car. He poked his head out of the window to get a better look and was zapped so hard that his glasses melted on his face.

## Now You See Us . . .

There have been lots of reports from all over the world of visits by strange-looking creatures and many of these have been witnessed by several people at the same time. One farm in Kentucky, USA, was dropped in on by a whole group of little egg-headed men with bright yellowy-green eyes. When the police turned up, they disappeared, only to reappear once the police had gone.

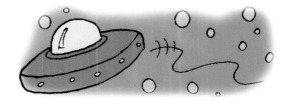

## Healing Powers

Other experiencers claim to have been healed by their encounter with ufonauts. One deputy sheriff from Texas, USA, had a serious bite heal up completely after a close encounter. A 73-year-old man in Argentina found that he had new teeth growing – very handy. Some people who have seen UFOs and made contact with their occupants have benefited greatly from the experience, becoming more creative and artistic.

## Disappearing Act

It is said that some extraterrestrials don't have solid bodies and can move in and out of their crafts without going through doors. Or that half of their bodies appear to be transparent, so that some seem to have no legs and to be floating in mid-air.

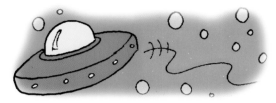

## Memory Loss

Many people who claim to have had an *encounter* (which means that they have been abducted or kidnapped by ufonauts and examined) find that they can't remember what happened to them but can recall the whole experience if they allow themselves to be hypnotized.

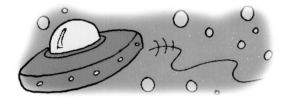

## Mind the Bend!

Some ufonaut spacecraft have been seen traveling at speeds estimated at around 1,600 km (1,000 miles) per hour. Incredibly, they seem to be able to change direction in a split second.

153

## Who's the Dummy?

In 1947, near Roswell army base in New Mexico, USA, a witness saw a spacecraft crash. When he went to look he saw several bodies lying on the ground. They were a bit like humans but smaller, with bigger heads and no hair. The army came along and closed off the area. Later the official explanation was that it was a crashed weather balloon and that the "ufonauts" were crash test dummies. One of the "dummies" was supposedly still alive.

## Beam Me Up, Scotty!

In 1975, a woodcutter in Phoenix, Arizona, USA, was sucked up into a beam of light while his workmates watched in horror. He returned six days later and remembered being taken into a flying saucer before he fainted, but he couldn't remember anything after that.

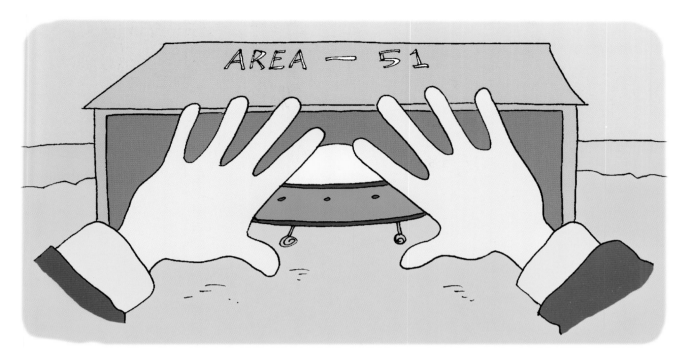

## Secret Location

Area 51, at Groom Lake, Nevada, USA, is where the US government does its UFO research, and any crashed or damaged UFOs are likely to be taken there for examination. But they don't like anyone snooping around – you'll soon be escorted out of the area or black helicopters will buzz you much too close for comfort. So someone is taking it seriously!

## Bulletproof?

Both ufonauts and their craft seem impervious to human weapons. One man who shot at, and hit, an extraterrestrial, said that it sounded like he had "shot into a bucket," and his target was unhurt. Anti-aircraft guns were fired at some UFOs over Los Angeles in 1942, and although they seemed to have been hit, they weren't damaged.

## False Gods

Do stories of gods and angels coming down from the heavens have anything to do with ancient sightings of extraterrestrials? Are the ancient gods merely beings from another planet who could do unearthly things and were therefore worshipped? A tribe in West Africa have cave paintings of their god Nommo, who told them things about the stars and planets that have only recently been proved to be true.

## Cosmic Chuckle

One thing is for sure, if there are ufonauts out there (and why shouldn't there be?) they must find the earthlings' attempts to recreate space travel hilarious. *Star Wars* is probably their most watched *comedy* show.

## Where in the World?

Brazil is the country where most UFOs are seen and several encounters have taken place. There are also places in the world that are known as "windows," where a lot of UFO activity is seen. Some prehistoric sites are connected by straight lines called *ley lines*, which are lines of power or force along which special sites such as stone circles, burial mounds, churches and so on have been built. These sites, and the ley lines on which they sit, are supposed to be "windows."

# Fieldwork

Crop circles are another funny phenomenon. They are large and sometimes complex shapes made by flattening a crop (usually wheat or barley). We still don't know how they're made because there are usually no marks to show where the "cropper" got in or out.

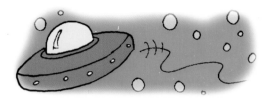

In July 1990, one of the most wonderful crop configurations yet appeared at Pewsey in Wiltshire, UK. Over 80 meters (260 ft) long, this was the first in a series of "double pictograms" and boasted ringed circles with "satellites" at the top of the formation. The farmer charged an entrance fee to the thousands of visitors who flocked to see them.

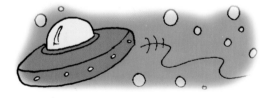

Crop circle experts claim that they can always tell which of the circles are faked. Most fakes are ragged, with lots of broken plant stalks.

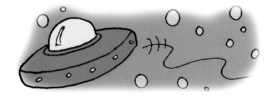

In the USA a huge crop pattern at Highland, Kansas, was claimed to be a prank. But someone was taking it seriously – the mysterious government investigators who sealed off the area tried to hide the circle and prevented the item from appearing on the local news.

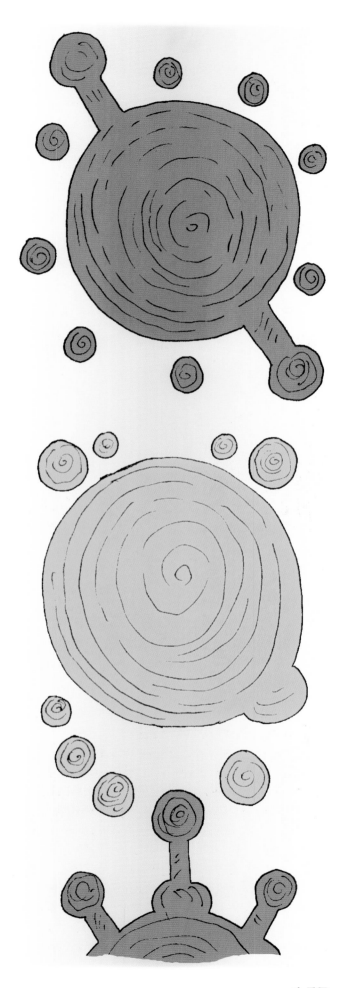

157

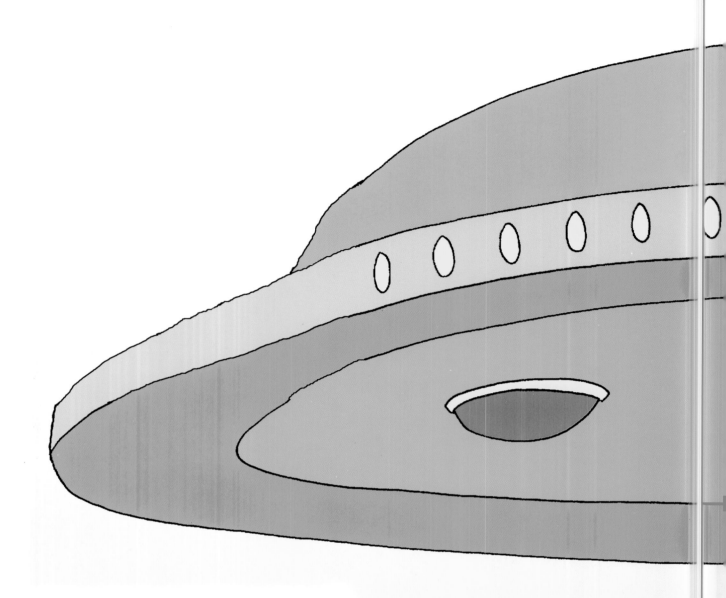

## Hoax Hoaxers

In 1991, a couple of British men set out to prove that crop circles could be made by humans and not extraterrestrials. In fact, they admitted that they were responsible for most of the circles around the country.

They claimed to have made the circles using a plank of wood attached to a central stake. The plank was then pulled round the stake by ropes or chains so that it flattened the corn in a circular pattern. However, no matter how hard anyone tried to do this, circles made in this way bruised and broke the stalks of the corn. And how did they get into the field without leaving marks, or manage never to be seen by the farmer or anyone else? So the question is, were the hoaxers real hoaxers or was their hoaxing a hoax?!

**158**

159

# WACKY SUPERNATURAL WORLD QUIZ

Did you enjoy that little collection? Take a minute to look at this little quiz. See how many of the questions you can answer and then try it out on your friends. I bet they won't believe that any of the answers can be right. You and I know different though, don't we!

1. **Psychokinesis is the ability to:**

a  Make objects disappear?

b  Make objects move?

c  Ride a bicycle with your knees?

2. **A psychometrist can:**

a  Find lost objects?

b  Tell who owned an object?

c  Only cycle for 1 meter?

**3. At Lourdes there is a:**

a Healing statue?

b Healing spring?

c Nice burger restaurant?

**4. In 1944 a shower of what fell on the village of Hopwas:**

a Fish?

b Frogs?

c Plastic ducks?

**5. The aurora borealis is:**

a Lights in the northern sky?

b A kind of lightning?

c A boring girl called Aurora?

**6. Kenneth Arnold was the first man to see:**

a Aliens?

b Flying saucers?

c The Queen in a bathing suit?

Answers on page 173.

# WACKY
# DOUBLE TAKES

## Any Post?

A young man was standing outside a shop in London, UK, several miles from his home, when a stranger with an American accent asked him if he could help to find an address that the American was unable to find. The American showed the man a letter that he had been given by friends in the States to deliver when he arrived in England. The young Englishman looked at it, and guess what? It was addressed to him!

## Same Scene

In 1942 a man who was stationed at Taversham Hall in England ordered a second-hand copy of a music book from Foyle's bookshop, London. When the book arrived in the post, the man opened it while looking out of the window of Taversham Hall. As he opened the book an old postcard dropped out. It was dated 1913, but the sepia photograph was of exactly the scene that the man was looking at out of the window.

## Double-Quick Minds

Twin brothers George and Charles can remember what the weather was like on every day of their lives. They can also work out the day of the week for any date up to 40,000 years into the future and once, when a box of matches spilled open, they both said there were 111 before the matches hit the floor.

## One Too Many

In primitive cultures twins were thought of as being strange, because it was believed that they shared the same soul. Often, one or the other would be killed at birth. It's thought that the historical "Man in the Iron Mask" might have been a twin who threatened his other "half." The man was kept in a iron mask which couldn't be removed, and imprisoned for life, although he was treated well. He is thought to have been King Louis IX of France's identical twin, who had to be kept locked away so that he couldn't claim the throne. Great thing to do to your brother, eh?

## Seeing Double

In 1867 the Reverend Charles Bingham was in a museum in Paris when he realized that another "man of the Church" was giving him odd looks. The Reverend thought that he might know him so he introduced himself – at the same time as the other man introduced *him*self. They were both named Reverend Charles Bingham.

Two men had a car accident. When the policeman asked each of them their name, both gave the same answer. Not surprisingly, the policeman told them to stop mucking around, but it was true. They were both called Ian Jones.

## Separated at Birth

Twin sisters Dorothy and Bridget were born and separated at birth, and only reunited when they were 34 years old. When they did meet, there were an incredible number of similarities between them (apart from the fact that they both looked the same!).

They both had one son: one was called Andrew Richard, the other, Richard Andrew. Each of them wore exactly the same number of rings and bracelets on each hand and wrist. Both had kept diaries in the year 1960 but no other year. The diaries were the same make and color, and the days for which no entries were made were the same in both diaries. Both loved historical novels and had the same favorite author.

## Mexico

Mexico City is the largest city in the world and over 20 million people live there. It is so polluted that breathing the air there on a normal day is like smoking two packs of cigarettes. This is due to the high altitude (therefore less oxygen) and thermal inversion, which means that the layer of smog traps the air below, so that there is even less oxygen.

You're very likely to be robbed in Mexico City – in fact, it's probably cheaper to take a taxi than a bus, because you're more likely to be pickpocketed on one of the crowded buses so a taxi will actually cost you less!

## Bearded Wonder

Queen Elizabeth I put a tax on all Englishmen with beards so it rapidly became fashionable to be clean-shaven. Up until the Middle Ages women with facial hair or hairy warts were thought to be witches and were ducked in water before being burned at the stake.

## Wacky Werewolves

A human who can transform himself into an animal is generally known as a werewolf. There's a story that a man saw his friend turn into a large wolf. The man fatally injured the wolf, then discovered that his friend was dying of a wound in exactly the same place.

## Motorcycle Madness

The Army Corps of Brasilia, Brazil, managed to get 47 men on one 1200cc Harley Davidson – and to ride it!

# Wacky Vampires

Corpses that rise from their graves at night and suck the blood of the living – Satanic manifestation or extraterrestrial beings? Anyway, all you need to ward them off is a string of garlic round your neck at night.

# Walking Dead

A *zombie* is a corpse without a soul or a will, animated by secret spells and potions. Zombies originated in Haiti, where the cult of voodoo abounds.

Here is one true zombie story – a young girl was supposed to have died and was buried in 1935, but three years later some of her friends saw her working in a shop. Though the girl moved stiffly and had a blank, "zombie-like" look on her face, her friends were sure it was her. Her grave was opened and it was found that the girl's body was missing from its coffin. There were rumors that the body had been stolen by a voodoo priest – the priest had since died, but his wife confirmed that he had made a zombie of the girl. She was sent to a convent in France.

**168**

## Double Decker
The huge River Nile, which flows northwards from central Africa through Egypt to the Mediterranean Sea, actually has another river flowing deep in the ground underneath it. This underground river holds six times more water than the Nile.

## Woodlouse Thermidor
Woodlice are ten a penny in most gardens, but they are crustaceans just like lobster, shrimp, prawn and crabs, so you could eat them if you wanted (boil first and toss in a little mayonnaise).

## Lake Eyre 1964
Donald Campbell chose this flat salt lake in Australia to race his car, Bluebird, to a new speed record. During the second run a tire was damaged and nearly sent Campbell to his death. The support team changed the wheel but were afraid that Donald Campbell would have lost his nerve after such a frightening experience. When they looked through the canopy he seemed perfectly calm but was staring at the screen. He went ahead with another run and broke the record. His mechanic later asked him what he'd been looking at during the wheel change. Campbell admitted he'd been looking at an image of his dead father that had appeared in the screen. It had told him it would be all right and given him the courage to continue.

# FINAL WACKY QUIZ

Did you enjoy that little collection? Take a minute to look at this little quiz. See how many of the questions you can answer and then try it out on your friends. I bet they won't believe that any of the answers can be right. You and I know different though, don't we!

1. **The man in the iron mask may have been:**

a  King Louis IX's twin brother?

b  King Louis IX's twin sister?

c  A really, really ugly guy?

SMASH!

2. **The largest city in the world is:**

a  New York?

b  Mexico City?

c  Immense-city!?

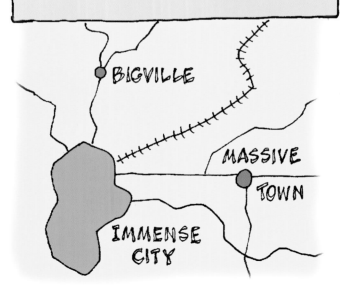

BIGVILLE

MASSIVE TOWN

IMMENSE CITY

**3. Queen Elizabeth I of England put a tax on:**

a  Hats?

b  Beards?

c  Farting?

**4. Woodlice are:**

a  Arachnids?

b  Crustaceans?

c  Nice on toast?

**5. A zombie is:**

a  A soulless animated corpse?

b  A kind of vampire?

c  A kind of sandwich?

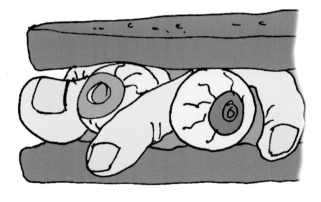

**6. Donald Campbell's record-breaking car was called:**

a  Bluethunder?

b  Bluebird?

c  Pixie?

Answers on page 173.

171

# ANSWERS

Did you have a go at all the quizzes? If not, go back and find them right now, you don't know what fun you are missing! Anyway, here are all the answers, but no cheating!!

## Ancient World Quiz

1. Edgar Cayce believed that the people of Atlantis had television sets.
2. Tutankhamun was a pharaoh of ancient Egypt.
3. The Great Pyramid at Giza was constructed from two million blocks of stone.
4. The Sphinx has the body of a lion and the head of a man.
5. The legendary lake of El Dorado was said to be full of gold.
6. The largest stone circle in Great Britain is Avebury.

## Natural World Quiz

1. Earthquakes are caused by shifting crustal plates.
2. A volcano that no longer erupts is known as extinct.
3. The largest rainforest in the world is found in South America.
4. Most of the water on the planet Earth is 3,000 million years old.

5. The Himalayan mountains are tall because they are young mountains.
6. The highest waterfall in the world is the Angel Falls.

## World of Animals Quiz

1. When a cat died in ancient Egypt its owner would shave their eyebrows.
2. Dog days are the hottest days of the year.
3. A quagga was an animal related to zebras.
4. The world's most poisonous scorpion is known as the fat-tailed scorpion.
5. The giant squid is related to the garden snail.
6. The stonefish is the most poisonous fish in the world.

## World of Humans Quiz

1. Menelek II of Abyssinia died after eating a Bible.
2. The science of iridology is concerned with the eyes.

3. The richest woman in the world is thought to be Queen Elizabeth II of England.

4. The longest swim ever took 742 hours.

5. The longest motorbike in the world is 7.6 meters long.

6. The Bishop of Rochester's cook was boiled to death for poisoning two of the Bishop's staff.

## Another Wacky Quiz

1. The largest cookie ever made was almost 25 meters in diameter.

2. The longest sausage ever made was just over 46 kilometers long.

3. It's bad luck to leave a knife stuck in a loaf of bread.

4. REM stands for Rapid Eye Movement.

5. If you want to be tall you should sleep more.

6. Nightmares are caused by worries.

## Monsters Quiz

1. Three-quarters of the Earth's surface is covered by water.

2. Cryptozoology is the study of undiscovered animals.

3. Giant squid are known to like the color red.

4. Loch Ness is in Scotland.

5. A mermaid has the body of a woman and the tail of a fish.

6. A sturgeon caught in the river Volga was 22 meters long.

## Ship Stories Quiz

1. The Flying Dutchman is a ghost ship doomed to sail the seas forever.

2. The captain of HMS Friday was named Friday.

3. The Philadelphia experiment was performed on the USS Eldridge.

4. The Bermuda Triangle is located in the Atlantic Ocean.

5. The Bermuda Triangle was first named in 1945.

6. A possible explanation for the Bermuda Triangle disappearances is waterspouts.

## Supernatural Quiz

1. Psychokinesis is the ability to make objects move.

2. A psychometrist can tell who owned an object.

3. At Lourdes there is a healing spring.

4. In 1944 a shower of frogs fell on the village of Hopwas.

5. The aurora borealis is lights in the northern sky.

6. Kenneth Arnold was the first man to see flying saucers.

## Final Quiz

1. The man in the iron mask may have been king Louis IX's twin brother.

2. The largest city in the world is Mexico City.

3. Queen Elizabeth I of England put a tax on beards.

4. Wood lice are crustaceans.

5. A zombie is a soulless animated corpse.

6. Donald Campbell's record-breaking car was called Bluebird.